Managing Risk in Projects

Projects are risky undertakings, and risk management is recognised as an integral part of managing the project. *Managing Risk in Projects* places risk management in its proper context in the world of project management and beyond, emphasising the central concepts essential to understanding why and how risk management matters, and presenting proven practical approaches to addressing risk in any project.

The risk management world has changed significantly since the first edition, with advances in risk management practice reflected by changes in international standards and guidelines, as well as significant developments in their implementation. This second edition reflects these changes, and has been completely updated to address progress in the practical application of risk management to projects. Two new chapters have been added, the first discussing how to manage risk in complex projects, and the second considering the role and influence of risk leaders outside the project arena in setting the context and environment for successful risk management. New material also addresses enterprise risk management and risky decision-making.

Throughout, the book offers a concise description of current best practice in project risk management whilst introducing the latest developments, to enable project managers, project sponsors and others responsible for managing risk on projects to do just that – effectively.

David Hillson, known globally as 'The Risk Doctor', is a recognised thought leader in risk management. Dr Hillson has received multiple awards and many of his innovations are now standard practice.

T0262807

"Once again David has clearly highlighted the importance of Project Risk Management in creating value for an enterprise's stakeholders. This edition's introduction of risk management in complex projects, and the importance of risk leadership, maintains this book's place as the cornerstone of any project risk management professional's resource library."

Wes Cadby, *CFIRM, Head of Risk Management,*
Sellafield Limited, UK

"David Hillson's classic 2009 book *Managing Risk in Projects* accelerated the shift of risk to centre stage in the project management field. This new update does it again, with the additional attention on complexity in projects, governance and strategic relevance. With rapidly increasing volatility and uncertainty in the world, including the growth in AI and machine learning, threats and opportunities are everywhere. If you don't want your critical projects to go up in flames, read this new book. Ignore it at your own risk!"

David L. Pells, *PMI Fellow, HonFAPM, Managing Editor,*
PM World Journal, USA

"This book introduces fresh and innovative ideas for effectively managing risk in projects, programmes, and portfolios. This latest edition provides valuable insights that are directly applicable to risk management in lean and agile environments. Dr Hillson continues his tradition of delivering accessible content that emphasises not only the practical application of risk management tools but also a deep understanding of the concept of risk in various organisational activities."

Bruno Rafael D. de Lucena, *PhD, Professor, Masters in Administration,*
Federal University of Pará, Brazil

"This second edition of David's book includes useful new material that addresses a wide range of recent changes in the evolving landscape of risk and projects. It is a succinct book that belies the breadth of topics it covers. While the focus is generally high-level, it contains a wealth of detail that will allow executives and project managers to tailor their risk management activities to the particular circumstances of their projects and the kinds of uncertainties they face. This book is written in David's typically clear language that is easy to read and understand. I strongly recommend it."

Dale Cooper, *PhD, CEO, Broadleaf Capital International, Australia*

"*Managing Risk in Projects* has been an excellent source of practical information on project risk management for the past 15 years. This second edition builds on that with expanded coverage of Enterprise Risk Management (ERM). Widening the description of project risk management to include programmes, portfolios, and the enterprise provides a much-needed strategic view of managing risk. David also provides insight and practical advice on addressing risk in complex projects, broadening our thinking on risk management and providing new contextual depth to the topic."

Cyndi Snyder Dionisio, *MBA, PMI-RMP, Practice Director,*
IIL Inc., New York, USA

Managing Risk in Projects

Second Edition

David Hillson

Routledge
Taylor & Francis Group

LONDON AND NEW YORK

Designed cover image: Getty Images / CasarsaGuru

Second edition published 2025
by Routledge
4 Park Square, Milton Park, Abingdon, Oxon, OX14 4RN

and by Routledge
605 Third Avenue, New York, NY 10158
Routledge is an imprint of the Taylor & Francis Group, an informa business

First edition published by Gower 2009 and Routledge 2016

British Library Cataloguing-in-Publication Data
A catalogue record for this book is available from the British Library

Library of Congress Cataloging-in-Publication Data
Names: Hillson, David, 1955- author.
Title: Managing risk in projects / David Hillson.
Description: Second edition. | Abingdon, Oxon; New York, NY: Routledge, 2024. | Includes bibliographical references and index.
Identifiers: LCCN 2023055424 (print) | LCCN 2023055425 (ebook) | ISBN 9781032557281 (hardback) | ISBN 9781032557298 (paperback) | ISBN 9781003431954 (ebook)
Subjects: LCSH: Risk management. | Project management.
Classification: LCC HD61 .H476 2024 (print) | LCC HD61 (ebook) | DDC 658.4/04--dc23/eng/20231130

LC record available at https://lccn.loc.gov/2023055424
LC ebook record available at https://lccn.loc.gov/2023055425

ISBN: 978-1-032-55728-1 (hbk)
ISBN: 978-1-032-55729-8 (pbk)
ISBN: 978-1-003-43195-4 (ebk)

DOI: 10.4324/9781003431954

Typeset in Garamond
by Deanta Global Publishing Services, Chennai, India

Contents

List of Figures   viii

List of Tables   x

Foreword to the First Edition   xi

Foreword to the Second Edition   xiv

Preface to the First Edition   xviii

Preface to the Second Edition   xx

1 Uncertainty and Risk   1

Current Sources of Uncertainty   1

Responding to Uncertainty   3

Distinguishing Between Uncertainty and Risk   5

A Pragmatic Distinction   7

Three Refinements   8

All Risks Are Uncertain, but Not All Uncertainties Are Risks   11

2 Risk and Projects   13

What's Wrong With Projects?   13

Why Are Projects Risky?   16

Why Manage Risk in Projects?   20

Individual Risks or Overall Risk (or Both)?   20

Why Is Risk Management Important to Projects?   22

3 Managing Risk in Practice   25

Towards a Risk Management Process   26

From Narrative to Reality   29

Describing the Risk Process 31
Not 'One-Size-Fits-All' 54
More than a Process 58

4 Managing Risk in Complex Projects **59**

The Challenge of Complexity 59
Identifying Risk in Complex Projects: Futures Thinking 63
Managing Risk in Complex Projects: Adaptive Resilience 68
Practical Project Continuity Management 70
Current Guidance on Risk and Complexity 73

5 Risk and People **75**

Understanding Risk Attitude 76
Influences on the Risk Attitude Spectrum 80
Risk Attitudes and the Risk Process 81
Risk and Decision-Making 83
Managing Risk Attitudes 86
People Plus Process 88

6 Integrating Risk Management With Wider Project Management **89**

Managing Risk Throughout the Project Lifecycle 90
Contribution of Risk Management to Other Project
Management Disciplines 96
'Built-In, Not Bolt-On' 98

7 The Bigger Risk Picture **99**

Strategy, Tactics and Projects 99
Hierarchy of Objectives, Hierarchy of Risk 102
Managing Enterprise Risk Across Boundaries 104
Effective Enterprise Risk Management 107
Project Risk Management in the Wider Risk Context 109
Managing Risk in Programmes and Portfolios 109
Remaining Challenges 112

8   Risk Leadership   **114**

Leadership Versus Management   115
Who Are Risk Leaders and What Do They Do?   117
Behaviours of Effective Risk Leaders   123
Risk Leaders – An Essential Contributor to Risk Management   125

9   Sustainable Risk Management   **127**

Risk Energetics   128
Internal Factors   130
External Factors   132
Risk Energetics Across the Project Lifecycle and Beyond   135
Proving It Works   137
Why Bother?   138
And Finally...   141

References and Further Reading   **143**

Index   **146**

Figures

Figure 1.1   Old World–New World   4

Figure 2.1   Linking Projects to Strategy   15

Figure 2.2   Standish CHAOS Data on Project Success 2011–2015   15

Figure 2.3   Relationship Between Risk and Reward/Loss (Indicative)   19

Figure 2.4   Risk Management as a CSF for Project Success   24

Figure 3.1   Risk Process   31

Figure 3.2   Double Probability-Impact Matrix   42

Figure 3.3   Example Probability-Impact Scoring Scheme   42

Figure 3.4   Example Bubble Diagram   43

Figure 3.5   Example Risk Prioritisation Chart   44

Figure 3.6   Example S-Curve From Monte Carlo Analysis   47

Figure 3.7   Overlaid S-Curves   48

Figure 4.1   The Cynefin Framework   60

Figure 4.2   Range of Possible Futures (Before Action)   67

Figure 4.3   Range of Possible Futures (After Action)   67

Figure 5.1   The Risk Attitude Spectrum   78

Figure 5.2   The Triple Strand of Influences on Risk Attitude   80

Figure 5.3   Making a Risky and Important Decision   84

Figure 5.4   The Seven As Model for Managing Risk Attitude   86

Figure 6.1 Alternative Views of Project Start and End-Points   91

Figure 7.1 Strategy and Tactics   100

Figure 7.2 The Organisation as a Hierarchy of Objectives   101

Figure 7.3 Sources of Risks at Intermediate Levels   105

Figure 7.4 Example Risk Efficiency Graph   111

Figure 8.1 Risk Leadership Areas of Responsibility   116

Figure 8.2 A-B-C Model   118

Figure 9.1 Risk Energetics – Decay and Damped Curves   128

Figure 9.2 Risk energetics – Desired Curve   130

Figure 9.3 Risk Energetics – Updates and Reviews   136

Figure 9.4 Risk Energetics – Rising Trend   137

Tables

Table 1.1   Dictionary and Thesaurus Definitions of Uncertainty and Risk   6

Table 1.2   Definitions of 'Risk' as 'Uncertainty That Matters'   9

Table 2.1   Definitions of 'Project'   14

Table 2.2   'Risks' vs. 'Risk' in Project Risk Management Guidelines   22

Table 3.1   Informal and Formal Risk Process Steps   30

Table 3.2   Mapping Generic Risk Process to Risk Standards   32

Table 3.3   Defining Terms for Probability and Impacts to Reflect Project Risk Thresholds   35

Table 3.4   Example Risk Breakdown Structure (RBS)   36

Table 3.5   Risk Management Plan Sample Contents List   37

Table 3.6   Typical Risk Register Data   40

Table 3.7   Roles and Responsibilities Within the Risk Process   56

Table 5.1   Risk Attitude Definitions and Characteristics   79

Table 5.2   Influence of Risk Attitude on Key Points in Risk Process   82

Table 8.1   Leadership and Management   115

Table 9.1   Benefits of Risk Management   140

Table 9.2   Cost-Benefit Analysis for Project Risk Management   141

Foreword to the First Edition

Simone Wray

When I started in risk management – coming on for over 15 years now – what we called 'risk management' was in reality the management of *insurable risk*, mainly through insurance, while project managers had an established set of tools to identify and manage *project risk*. Both groups of people knew there should be synergy, but there it tended to stop. Worse, the separation created an element of competition between the worlds of project management and risk management. It was not unknown for project managers and risk managers in the same organisations to have no real contact – or even purposefully avoid each other. In some cases, they attempted to recognise each other's contribution to the management of risk within the organisation, but failed to see how to make a real connection between their respective roles.

As risk management and related disciplines such as internal audit and business continuity evolved, more territorial struggles followed. Business continuity managers felt that risk management was their domain. Following the introduction of *The Combined Code on Corporate Governance* in 1998, audit managers saw risk management coming within their ownership.

But the development of risk management has also led to an appreciation of the need to adopt a consistent and planned approach to the management of all risk – a so-called enterprise risk management approach. Enterprise risk management is a concept that embraces the management of all business risk across an organisation. It has however only been recognised comparatively recently as something that can add value for an organisation by providing effective business tools to manage risk.

This probably has something to do with how the concept of risk has evolved: from the initial idea of an event which was inevitably negative and could damage operations, through to a broader understanding that risk reflects uncertainty which can have a detrimental or positive effect on strategic objectives. The first chapter of this book explores this connection between risk and uncertainty in a very simple way. Reading this chapter alone will clarify a lot of unresolved thoughts and debates among the risk, project and other related communities.

Now we find that risk management has created a connection between the project manager and the risk manager because it provides a common language for dealing with uncertainty. In fact, it enables all professionals from different functions to communicate better with each other on the subject of risk – and since most projects bring a range of professionals together, this leads to more effective management of risk within the project.

Risk managers and project managers need to be professional best friends. Working in separate towers will only lead to frustration for each of them, while if they understand each other's roles and support each other's purpose, the result should be a win-win situation. Both are focused on the success of the organisation that they work for, and collaboration in the effective management of risk is a great contribution to organisational success. This must surely be beneficial at a personal level, as well as in today's working environment where increasing importance is placed on being able to demonstrate the difference that you are making to value and that you can work as part of a team.

To be successful in delivering the benefits it envisages to its stakeholders, an organisation needs a coherent, aligned and hierarchical set of objectives that provides a common thread from the strategic level to tactical delivery. Having established these objectives, the organisation needs to achieve them, despite uncertain operating environments. Projects do not exist in isolation within an organisation; they are one of the ways by which organisations make their intentions material.

To use David's words, risk is 'uncertainty that matters' – from whatever source. To overcome any conflict between the management of risk at project level and at strategic level, an enterprise risk management approach ensures that risk is managed consistently at all levels of the organisation across the hierarchy of objectives. Otherwise, important risks that occur in the gaps or that result from correlation between apparently separate exposures will be overlooked or ignored. This is certainly the case if business risk and project risk are identified and managed in isolation.

Done in this way, enterprise risk management offers an integrative framework for the business that leads to successful project delivery and ultimately to realisation of strategic benefits and value. There is a bigger picture, and David explores this in this book. The contribution of project risk management to this overall success requires it to be integrated fully into the wider hierarchy of enterprise risk management, with particular attention to the interface with the next level up, namely programme risk management. Only then can project risk management play its full part in delivering value to the organisation.

Finally let's turn to behaviour. A project manager needs to understand their own influence and that of the project team on the response to uncertainty resulting from attitudes to risk. *Managing Risk in Projects* explores what often can be a missing link in such reference texts, the important behavioural side of the people involved in the project and that of the organisation itself in terms of its culture and ability to learn. Wherever there are people there is risk, and no organisation would exist without people.

This book sets out to discover why risk management is important in the context of projects, how it should be implemented, how risk outputs should be used both within and outside the project, and what is necessary to maximise risk management effectiveness. For newcomers to the project or risk professions, it provides a practical overview of risk management practice within the specific context of projects, and how this relates to enterprise risk management. More experienced project managers and risk managers should question developed thinking and practice from time to time and, as David mentions in his Preface, they may find themselves rehearsing first principles in order to develop their take on innovation and best practice. *Managing Risk in Projects* provides a fast track to both. Essential reading for either audience, the book takes current thinking in risk management and creates the necessary links to show the possibility of a joined-up approach. The important point for all readers, whatever their level of experience, is to take the key messages from each chapter and consider how to apply them within the context of their own organisations.

Simone Wray
Chairman, Institute of Risk Management
February 2009

Foreword to the Second Edition

Dale Cooper

The international standard IEC 62198 *Managing risk in projects – Application guidelines* sets out principles for effective, efficient and consistent project risk management. The first of these principles is 'Risk management creates and protects value.' The standard expands on this principle:

> Risk management contributes to demonstrable progress towards organizational objectives and improvement of performance and quality in projects and the assets, products and services they create. The objectives shall be understood clearly by all parties.

This first principle is related directly to David Hillson's pragmatic definition of risk as 'uncertainty that matters'. Risk is characterised by uncertainty – if there is no uncertainty there can be no risk – and the effects of that uncertainty on the objectives, outcomes and value that an organisation derives from undertaking a project.

The second principle in IEC 62198 is 'Risk management is part of decision-making,' with the further commentary that:

> Risk management helps decision makers make informed choices about the project, within each stage of its life, prioritize actions and distinguish among alternative courses of action. This implies that all decisions should consider risk.

This new edition of David's book provides a concise view of current 'good practice' in project risk management, with a consistent focus on making sound and justifiable decisions that add value and generate better outcomes. He takes a high-level view, suitable for all project risk practitioners, that allows many general and wide-ranging aspects of project risk management to be discussed, including topics that are often omitted or treated only in a limited way in other books.

Three aspects are of particular interest: how projects fit in with an organisation's purpose and objectives, how complexity affects project risk management, and the importance of people in making project risk management work well. These topics and their implications are discussed throughout the book.

Projects are not isolated, independent pursuits. They are often part of programmes or portfolios of organisational endeavour, usually developed and implemented as vehicles for achieving strategic and business outcomes. When viewed in this way, project risk management has many features in common with enterprise risk management. Similar processes and tools can (and should) be applied, so uncertainty is addressed in a consistent way across the organisation.

Setting projects into their organisational context has several implications.

- Projects start as soon as a business need is identified. Project management begins well before the delivery phase, and so must project risk management.
- The objectives for a project are rarely as simple as good delivery: completing the project on time, within budget and to an acceptable standard of quality. They usually include much wider requirements that reflect the business need.
- The effects of uncertainty on objectives can be positive, negative or both. Project risk management must focus on capturing opportunities as well as reducing threats.
- It is important to distinguish between individual *risks*, usually of interest within a project, and the overall *risk* of a project, usually of interest at an organisational level. This requires processes that facilitate allocating priorities to risks, so project managers can make better decisions, and that also support a sensible integrated view of project risk for executive decision makers.

The complexity of an organisation's environment and the projects in which it engages influences how its projects are implemented. Increasing complexity has encouraged a move from linear and waterfall methods towards agile approaches to project management. David includes a chapter in which he discusses volatility, uncertainty, complexity and ambiguity, the way organisations adjust what they do to remain resilient in these circumstances, and the implications for how they manage risks.

This is an important discussion because there is little consensus about the best way to address risks in complex projects. Traditional, common-practice risk management processes alone are rarely sufficient. They must be adjusted and augmented, initially with techniques from other forward-looking disciplines such as horizon scanning, scenario planning and futures planning that support better risk identification, and then with approaches directed towards improving continuity, adaptability and resilience. Comprehensive monitoring and review processes are likely to be needed too, to support proactive identification of trends and changes that indicate when new threats and opportunities may be emerging, in a project environment that is evolving in unanticipated ways.

A strength of this book is the recognition throughout that project risk management is not a theoretical process. It must be implemented by people, and people are central to its success in practice. Start with the notion of risk as 'uncertainty that matters': while business objectives provide the main guide to *what* is important, it is people, with their individual attitudes to risk and their personal risk appetites, who determine *how much* a source of uncertainty matters and the priority it should be allocated when making decisions.

Attitudes to risk are intertwined with and influence many aspects of projects and the organisations that undertake them. The attitudes to risk of executives and project managers, and the leadership these people provide, create and shape a culture that is appropriate for the risks the project faces. This in turn drives the enthusiasm and energy project teams devote to understanding uncertainty and managing risk in their roles, and the way they integrate risk-related thinking into their day-to-day project activities.

This reinforces IEC 62198's third principle: 'Risk management is an integral part of all organisational processes associated with a project.'

> Risk management is not a stand-alone activity that is separate from the main activities and processes of the project or the organization. Risk management is part of the responsibilities of

project managers and of staff at all levels. It is an integral part of all the organizational processes associated with a project, including strategic project and investment planning, project management and management of project change.

This second edition of David's book includes useful new material that addresses a wide range of recent changes in the evolving landscape of risk and projects. It is a succinct book whose length belies the breadth of topics it covers. While the focus is generally high-level, it contains a wealth of detail that will allow executives and project managers to tailor their risk management activities to the particular circumstances of their projects and the kinds of uncertainties they face.

This book is written in David's typically clear language that is easy to read and understand. I strongly recommend it.

Dr Dale Cooper
Cammeray, New South Wales, Australia
November 2023

Preface to the First Edition

While it is very stimulating to be on the leading edge of any discipline, it can be a dangerous and lonely place sometimes. All pioneers need a safe base from which to set out on their adventures of exploration and discovery, and a home to which they can return. It is not possible for most of us to live permanently on the mountain heights or in the depths of the jungle, no matter how absorbing those places might be for a time. I view my relationship with risk management in a similar fashion. You would rightly expect the Risk Doctor to enjoy life on the edge, and certainly I find great fulfilment in working at the boundaries of our profession, seeking to develop new understanding and practical approaches to managing risk better. But I also find myself returning time and again to the fundamentals of our fascinating topic, rehearsing the first principles to ensure that any innovation is properly grounded in the essentials of risk management theory and practice.

That's why I'm pleased to offer this book covering the vital topic of *Managing Risk in Projects*. Projects are risky undertakings, for a number of reasons which are explored in the early chapters. As a result, modern approaches to managing projects have all recognised the central need to manage risk as an integral part of the project management discipline. Risk management is established as a core knowledge area and competence for project management practitioners, and there is wide consensus on what it entails. This book describes how risk management can be applied to all projects of all types and sizes, in all industries, in all countries. It places risk management in its proper context in the world of project management and beyond, and emphasises those central concepts which are essential to understand why and how risk management should be implemented on all projects. The generic approach detailed here is consistent with current international best practice and guidelines, but also introduces key developments in the risk management field, to ensure that readers are aware of recent thinking, focusing on their relevance to practical application.

This is of course what former British Prime Minister John Major would have called 'back to basics'. This book addresses the basics of risk management as implemented in the project context, with enough detail to explain why it is important, what is involved in implementing the risk process, and how to use risk-based outputs when managing projects. New leading-edge material is however also included, including the results of recent research on the effect of risk attitudes on decision-making, the interface between risk management at project-level and programme-level, and the ideas of 'risk energetics' as a framework for understanding the Critical Success Factors for effective risk management.

Throughout, the goal has been to offer a concise description of current best practice in project risk management while also introducing the latest relevant developments, to enable project managers, project sponsors and others responsible for managing risk in projects to do so effectively. While the presentation of the ideas in this book represents my own views of the subject, I have of course drawn on the wisdom and insights of many who have gone before. Unfortunately, they are too many to name individually, but they include the pioneers of project risk management who are well known to most.

I wish to acknowledge the support of my publisher Jonathan Norman from Gower Publishing, whose constant encouragement and enthusiasm makes me want to keep writing for him. My family and friends have also been patient and understanding, especially my wife Liz, who showed remarkable self-restraint when I suggested I should write another book on risk. I'm also grateful to my professional colleagues and clients who have been courteous enough to allow me to try out some of my ideas on them.

And finally, I offer this book to those who know that risk management is important to project success but aren't quite sure why, or who feel they could do it better if only they knew how, as well as all who are committed to managing risk in projects. By coming 'back to basics', we can ensure sound foundations which will allow us to build an effective approach to project risk management, leading to more successful projects and businesses. In these uncertain times, what more could we want?

Dr David Hillson, 'The Risk Doctor'
Petersfield, Hampshire, UK
January 2009

Preface to the Second Edition

Projects remain the primary means by which organisations create value for their stakeholders. But projects also remain as risky as they ever were, perhaps more so. That's why effectively managing risk is still a vital contributor to project success, and it's also why a book like *Managing Risk in Projects* is still relevant.

But why do we need a second edition? Much has changed in the risk management world in the past 15 years or so, since the first edition was published in 2009. This includes advances in risk management practice, reflected by matching changes in standards. Most significant has been update of the international standard ISO31000:2018 *Risk management – Guidelines*, and subsequent production of a family of related standards covering specific types of risk (including IEC/ISO31010:2019, ISO31022:2020, ISO31030:2021 and ISO31073:2022, with more in development). Other generic standards have also been revised to match the new ISO31000, including ISO21500:2021 *Project, programme and portfolio management – Context and concepts*. Still in the project arena, risk standards from professional bodies such as the Project Management Institute (PMI) and the UK Association for Project Management (APM) have been updated recently, extending risk guidance from projects into programmes and portfolios. All these developments in standards and guidelines are reflected in this second edition.

More importantly, the book has been completely updated to address developments in the practical application of risk management to projects, while retaining the broad coverage of the first edition. In particular, two new chapters have been added. The first discusses how to manage *risk in complex projects*, where traditional approaches to risk management have proved inadequate to deal with the emergent nature of risk in complexity. Secondly, a new chapter considers the role of *risk leaders* outside the project arena, whose vital influence in setting the context and environment for risk management has become increasingly recognised. As for the first edition, the aim has been to describe current best-practice in project risk management whilst introducing the latest developments.

On a personal note, I've been actively involved in the risk world for nearly 40 years, and the fascination remains as strong as ever. What keeps me engaged is the constantly changing risk landscape, with new risk challenges emerging on a regular basis, demanding innovative thinking and novel approaches to tackle them. My own involvement in risk management started with applying risk thinking and techniques to a troubled project, as a test case for my company to decide whether risk management was worth investing in. We turned the project round by uncovering and addressing the main sources of uncertainty, the project was a success, and my interest was captured. Since then, I've seen the value of risk thinking demonstrated in multiple settings well beyond the world of projects, within different types of organisations across the globe, including governments, multinational corporations, small businesses, aid organisations, charities and social enterprises. But I still find myself drawn to the challenge of managing risk in projects, where innovation and new thinking are a way of life. I hope the second edition of this concise book will provide inspiration and guidance to new generations of project professionals, pointing a better way towards increased project success through effectively managed risk.

Dr David Hillson, 'The Risk Doctor'
Petersfield, Hampshire, UK
October 2023

Chapter 1

Uncertainty and Risk

Current Sources of Uncertainty

There can be little doubt that we live in a world characterised by uncertainty. It was not always so, at least in some important aspects. While the natural environment has always been uncertain (earthquakes, volcanoes, hurricanes, floods and other so-called acts of God), the social environment in which we live has changed dramatically in many respects, particularly in the WEIRD nations (Western, Education, Industrialised, Rich, Democratic), and the old certainties of previous generations no longer exist. In living memory, as little as two or three generations ago, people lived in stable communities where they knew everyone else. Each person understood and (for the most part) accepted their position in society, and their relation to others. For most individuals, their job choices were prescribed by their family position, and the concept of 'career' was alien to many. The choice of marriage partners was limited and sometimes even absent. It was possible for the majority of people living in that society to look ahead for two, five, ten years or more, and predict with reasonable certainty where they would be living and what they would be doing. Boundaries were fixed, horizons were limited, and both were largely known, understood and accepted.

Even beyond the local community, there was stability in large areas of the world, reinforced by the international power bases of the British Empire and Commonwealth, the United States of America, NATO, and the USSR and Warsaw Pact. Technology was slow-moving, and business practices and structures remained largely stable, with business planning cycles typically looking ahead by five to ten years.

While these societal characteristics can still be found in some parts of the globe, it is not the case in the 'developed' world today. We are experiencing unprecedented volatility, with huge degrees of flexibility and choice in all levels of society, including families, local communities, businesses and nations. Individuals have very few fixed points, and the degrees of freedom and mobility for many have increased dramatically (though not for everyone, of course, since all advanced societies still have their underclasses). Asking someone where they think they might be in two, five or ten years is likely to be met with puzzlement – how could we know?

Technological change has quickened to a rapid pace, with inventions being widely adopted in a very short timescale. Some innovations have become all-pervasive to a degree where it is hard to imagine life without them (for example, accessible computing, the Internet, wireless connectivity, mobile telephony), but they have arrived very recently and the take-up time has been very short. It is almost impossible to predict where technology might go next, with the possible outputs of R&D departments resembling science fantasy rather than realistic products. The rapid rise of Artificial Intelligence (AI) is challenging many aspects of society, including government, regulators, businesses and the general public. The planning cycle for most organisations has reduced dramatically, with typical horizons of one or two years at maximum, and often less.

Other aspects of modern society are characterised by new types of uncertainty that did not previously exist, leading to new unpredictabilities. For example, disease patterns used to be well understood two or three generations ago, and today we have sophisticated models for many of these diseases. However, we now face previously unforeseen challenges from new types of pathogens that did not exist before, such as genetic hybrids or nanobiotechnology. Global pandemics have re-emerged as a reality after an absence of a century, and these are likely to remain part of the fabric of the modern world, with unexpected implications. For example, the COVID-19 pandemic resulted not only in health challenges across the world, but also produced global supply chain instability, demographic upheaval, and irreversible changes in the workplace.

Financial markets have experienced volatility on a massive scale, and continue to do so, with implications for ordinary people having mortgages, savings or pensions. International power blocs are fluid and emergent, with the old masters giving way to new challengers such as the BRIC economies of Brazil, Russia, India and China (or perhaps the CHIME countries of China, India and the Middle East gulf states). Other non-national or supra-national groupings are also influential on the world stage, including both ethnic groups and multinational corporations, competing with the nation-state. Terrorism remains a major concern for many, but geopolitical tensions have also increased recently. Armed conflict between nations has reappeared on the global stage, as well as many ongoing internal struggles within nations that amount to de facto civil war.

Despite high levels of public interest and input from the scientific community, the implications of climate change and global warming remain unclear, leading to uncertainties in the required scope and pace of adaptation.

This rapid rise in uncertainty in so many dimensions of modern life has led to a crisis of confidence, with many believing that the world (or at least their world) is both out of control and uncontrollable. New terms are entering the common discourse, including *polycrisis* (the simultaneous occurrence of several catastrophic events) and *permacrisis* (a period of extended uncertainty, particularly where this arises from the combination of several overlapping catastrophic events). The United Nations Office for Disaster Risk Reduction (UNDRR) has recently issued a study which defines a new category of 'exis-tential risk', which they define as 'the probability of a given event leading to either human extinction or the irreversible end of development' (Stauffer et al., 2023). This report has a particular focus on biotechnology and Artificial Intelligence as major drivers of accelerating change, raising concerns about deliberate or accidental release of a novel pathogen that causes a lethal pan-demic, and development of transformative AI that could lead to entirely new technological hazards.

Responding to Uncertainty

Previous societies have used religion, science and law in an attempt to impose predictability on the uncertainties they faced. These frameworks gave some sense of order and meaning to life, setting events in a wider context. Each provided an external authority which sat above and beyond the individual, family, community or nation. By referring to these, it was possible to treat the world as more certain than it might have been in reality, resulting in a degree of stability and contentment.

In today's post-modernist world such external sources of authority are challenged, and people are left to make their own sense of their surroundings as best they can. The drive for certainty seems to be inherent in human nature, and we look for it where perhaps it cannot be found. For example, the rise in government regulations designed to minimise risk is an indication of how citizens expect their rulers to protect them from uncertainty and its effects, instead of taking responsibility for their own lives and choices and recognising that uncertainty is inherent in life. We demand certainty and precision from our scientists and we complain when they are unable to quantify risks from sources such as mobile phones, genetically modified foods or climate change. This fails to acknowledge that science is based on hypothesis and experimentation, knowing that the current state of human knowledge is incomplete and provisional, only approximating to reality and truth.

In the business world, organisations seek to predict change and respond to it, but the pace of change is in danger of overtaking the rate of learning, as illustrated in Figure 1.1. In what Obeng (1997) calls the 'Old World',

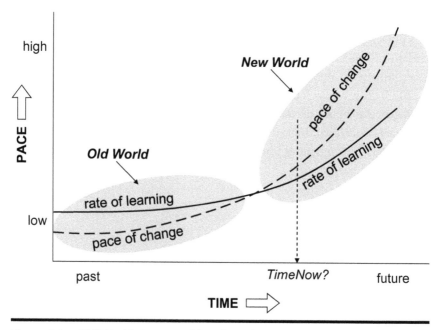

Figure 1.1 Old World–New World (Adapted From Obeng, 1997)

businesses were able to stay ahead of the curve by learning faster than their competitors and adapting to change as it occurred. In the 'New World' of rapid change, gaps appear as the ability of organisations to respond falls behind the pace of change. Here the winners will be those who are able to evolve and adapt, innovate and respond. A decade or two ago, we were arguably at the turning point between the two worlds, but now things have moved on and we have entered the New World (although the precise extent of disparity between pace of change and rate of learning is not clear). Businesses clearly need to change their paradigm in order to survive and prosper.

Clearly, some aspects of life today are more uncertain than ever before. This fact is inescapable. The only question is how we will cope with it. While individuals may implement a range of strategies for dealing with uncertainty, business looks to the discipline of risk management to address this question. In order to understand how risk management might meet the challenge of uncertainty, we first need to clarify the relationship between uncertainty and risk.

Distinguishing Between Uncertainty and Risk

If risk management is to help to tackle the challenges posed by an uncertain world, it must be properly focused and effectively implemented. This depends on having a definition of risk which is clear, unambiguous and widely accepted. The definition debate is not an abstruse irrelevance of interest only to academics and pedants. If we are unable to define a risk, we will not be able to undertake risk management effectively.

The first question is whether we need the word 'risk' at all. At first sight the terms 'uncertainty' and 'risk' seem similar. But how similar? Are they mere synonyms, able to be interchanged without confusion or loss of meaning? Or is there any real and useful distinction between the two?

Contrary to expectation, a dictionary or thesaurus will not help here (see Table 1.1). The disparate range of options for both terms does not support a clear understanding of their relationship. It seems that we need to look elsewhere to determine whether risk is the same as uncertainty.

Fortunately, others have already attempted to clarify a distinction between 'uncertainty' and 'risk' without resorting to a dictionary. Knight (1921) addressed this in the field of economics, separating insurable risk from true uncertainty. His approach drew on basic mathematical theory, that 'risk' arises from randomness with knowable probabilities, whereas

Table 1.1 Dictionary and Thesaurus Definitions of Uncertainty and Risk (Harpercollins, n.d.)

Term	Uncertainty	Risk
Dictionary	Lack of certainty; doubt; the state or condition of being uncertain; a state of doubt about the future or about what is the right thing to do.	The possibility of incurring misfortune or loss; hazard; the chance of injury, damage, or loss; dangerous chance.
Thesaurus	ambiguity; chanciness; changeableness; confusion; doubt; fickleness; hesitancy; inconclusiveness; indecision; precariousness; state of suspense; unpredictability; unreliability.	chance; danger; hazard; jeopardy; liability; likelihood; menace; odds; peril; possibility; probability; prospect; speculation; threat; uncertainty.

'uncertainty' reflects randomness with unknowable probabilities. The terms 'aleatoric' (from the Latin word *alea* meaning dice) and 'epistemic' (from the Greek word *episteme* meaning knowledge) are sometimes used to distinguish between these two. Decision-theorists take a similar approach, separating 'decisions under risk' where the probabilities of different outcomes are known (or at least knowable) from 'decisions under uncertainty' where probabilities are unknown (and maybe unknowable). Some philosophers suggest that as a result 'uncertainty' belongs to the subjective realm of belief, while 'risk' has an objective component based in fact or truth.

In theory this type of distinction may seem useful and clear, but in reality probabilities are rarely known with any precision or certainty. Throwing unbiased dice or flipping fair coins are idealised cases of risky situations, but any real-world example will not behave in so straightforward a manner. In most cases we cannot be sure that estimates of probability are correct, so even 'risk' is uncertain!

If we are to find a clear role for risk management in relation to meeting the challenge of uncertainty, discussions based in mathematics or philosophy are unlikely to yield usable solutions. A more pragmatic approach is required, which is useful in practice, and which supports effective risk management and good decision-making when conditions are not certain. Looking again at the definitions in Table 1.1, it appears that 'uncertainty' is a generic term, while 'risk' seems to be more specific. This may give a clue to how they may be usefully distinguished. Perhaps 'risk' can be seen as a subset or special case of 'uncertainty'.

A Pragmatic Distinction

Reviewing the world around us confirms that it is characterised by uncertainty in many forms arising from a variety of sources. However, the task of risk management is quite specific. It is to enable individuals, groups and organisations to make *appropriate decisions* in the light of the uncertainties that surround them. The key word here is 'appropriate'. How can we determine what response is appropriate for any particular uncertainty? One way is to separate the various uncertainties into two groups: those that matter to us, and those that do not matter. There are perhaps an infinite number of uncertainties in the universe but they do not matter equally; indeed, some do not matter at all while others are literally vital. As we seek to make sense of our uncertain environment and decide what to do in order to move forward, we need to know which uncertainties matter, and then respond appropriately to those. Any uncertainties which do not matter can be ignored, or perhaps reviewed from time to time to see whether they or our circumstances have changed to the point where they might now matter.

This leads to a proto-definition of 'risk' which offers a useful distinction to guide our thinking and practice:

'Risk' is 'uncertainty that matters'.

While this may not be suitable as a fully formed definition, it does point us in the right direction. Not every uncertainty is a risk, though risk is always uncertain. Risk becomes a subset of uncertainty, filtered on whether or not it matters. If risk management focuses on identifying and managing those uncertainties that matter, it will help us to respond appropriately. In fact, this is consistent with the earlier mathematical and philosophical distinctions between uncertainty and risk. For example, the outcome of a horse-race is usually uncertain, but unless an individual has bet on the result there is no risk for them. The uncertainty only becomes a risk when it matters, otherwise it is a mere intellectual curiosity or irrelevance.

To make this practical as a framework for risk management, we need to know how to decide whether a particular uncertainty matters or not. The key is to focus on *objectives*. These define what matters to any individual, group or organisation. Objective-setting is the process of describing our desired goal and the end-point that represents success. To concentrate on what matters means to link everything to achievement of agreed objectives. By defining 'risk' as that subset of uncertainties which matters, we are tightly coupling risk management to achievement of objectives, since the goal is to identify and manage any uncertainty that could affect our desired outcome. This provides a clear link between risk management and success, delivery, value and benefits. Where risks are effectively managed, the chances of achieving objectives will be optimised. Conversely, poor risk management will reduce the likelihood of success.

Making the link between risk and objectives moves us closer to a usable definition of risk. Risk is a type of uncertainty, but not every uncertainty is a risk. Instead, risk is that subset of uncertainty that matters, and we determine whether a particular uncertainty matters by considering the possibility that objectives might be affected. Of course, the uncertainty will only actually matter in practice if it occurs and becomes reality. So, our proto-definition of 'risk' as 'uncertainty that matters' can be expanded into:

> 'Risk' is 'uncertainty that, if it occurs, will affect achievement of objectives'.

Indeed, this form of definition is found in most of the current risk management standards and guidelines, as illustrated in Table 1.2. Each of the definitions shown in the table has two distinct parts: the first of these relates to some type of uncertainty, and the second part describes why it matters by linking the effect of the uncertainty to achievement of objectives.

Three Refinements

Before leaving the relationship between uncertainty and risk, three further important points arise.

Firstly, examination of Table 1.2 reveals an interesting detail in several of these definitions of risk, namely that risks are not wholly negative. This is explicitly stated in three of the listed definitions (Project Management Institute, 2019; European Commission Centre of Excellence in Project Management, 2018; Institution of Civil Engineers and Institute & Faculty of Actuaries, 2014), which use the phrase 'positive or negative' when describing possible impacts. The international risk standard ISO31000:2018 (International Organization for Standardization, 2018) maintains a neutral

Table 1.2 Definitions of 'Risk' as 'Uncertainty That Matters'

Source of definition	'Uncertainty...'	'...that matters'
ISO31000:2018 *Risk Management – Guidelines*. (International Organization for Standardization, 2018)	'Effect of *uncertainty*...'	'...on *objectives*.'
	'Note: An effect is a deviation from the expected. It can be positive, negative or both, and can address, create or result in opportunities and threats.'	
The Standard for Risk Management in Portfolios, Programs and Projects. (Project Management Institute, 2019)	'An *uncertain* event or condition...'	'...that, if it occurs, has a positive or negative *effect on one or more objectives*.'
APM Body of Knowledge, 7th Edition. (Association for Project Management, 2019)	'An *uncertain* event or set of circumstances...'	'...that would, if it occurred, have an *effect on achievement of one or more objectives*.'
	'The *potential* of a situation or event...'	'...to *impact on achievement of specific objectives*.'
Management of Risk [M_o_R4]: Guidance for Practitioners, 4th Edition. (Axelos, 2022)	'An *uncertain* event or set of events...'	'...that, should it occur, will have an *effect on the achievement of objectives*.'
PM² Project Management Methodology, Guide 3.0. (European Commission Centre of Excellence in Project Management, CoEPM², 2018)	'An *uncertain* event or set of events (positive or negative) ...'	'...that, should it occur, will have an *effect on the achievement of project objectives*.'
Risk Analysis and Management for Projects: A strategic framework for managing project risk and its financial implications, 3rd edition. (Institution of Civil Engineers and Institute & Faculty of Actuaries, 2014)	'A *possible* occurrence...'	'...which could *affect (positively or negatively) the achievement of the objectives* for the investment.'

definition of risk, but adds a footnote explaining that effects can be 'positive, negative or both'. If risk is 'uncertainty that matters', this is not about exclusively negative or adverse impacts on achievement of objectives. It is possible to imagine uncertain events or sets of circumstances which, if they were to occur, would be helpful towards achieving our goals. Such positive possibilities are usually called 'opportunities'. We might view these merely as 'good luck', the unlooked-for fortuitous events that could save time, save money, increase productivity, enhance reputation and so on. But current standards include opportunities in the scope of their definition of 'risk', as possible future events that might occur, and which if they were to occur would have an effect on achievement of objectives, and which therefore require us to identify and manage them proactively.

This double-sided concept of risk as threat and opportunity is not only present in the standards and guidelines, but it is increasingly being implemented in practice by leading organisations. There are a range of distinct benefits from adopting this wider approach to risk, including the following (Hillson, 2019):

- *Exploits more opportunities.* Instead of hoping to take advantage of any good luck that might occur, including opportunities explicitly in the risk process means that more of them will be identified in advance and managed proactively.
- *Permits trade-offs.* If the risk process concentrates only on identifying and minimising possible downside, it is likely that objectives will not be met. By finding and capturing opportunities and turning them into benefits or savings, some of the adverse effects of threats can be mitigated.
- *Increases chance of success.* A process which proactively seeks upside will inevitably deliver more successful outcomes, as at least some of the opportunities are captured.
- *Supports innovation and creativity.* The process of identifying opportunities requires a positive mindset which seeks improved ways to deliver value. This results in more innovative and creative thinking aimed at maximising results.
- *Increases efficiency.* It would be possible to implement a process for managing opportunities separately from the risk process. However, using a combined process to manage both threats and opportunities delivers synergies and efficiencies. These savings can be significant in an organisation which already has an established risk process used only

for threats, since the additional value can be obtained with minimal extra effort.

■ *Motivates teams.* People find it disheartening when risk is unmanaged and they are required to react to emerging crises or correct avoidable problems. A focus on upside risk with the potential for positive improvements in performance will increase motivation and job satisfaction.

A second key point arises from defining risk by linking uncertainty with objectives. This is the recognition that the typical organisation has a range of objectives at many levels. There are strategic objectives at the highest level, which are translated into increasing detail for implementation through the delivery elements of the organisation. Subsidiary objectives exist at financial, safety, regulatory, programme and project levels, among others. If risk is defined as 'uncertainty that, if it occurs, will affect achievement of objectives', then it applies wherever there are objectives. Risk management is not just relevant to technical or delivery disciplines, but affects every part of an organisation from top to bottom. Successful identification and management of uncertainties that matter is essential for success across the business at every level. This is why risk management deserves such wide attention, and we will return to this theme in Chapter 7.

Lastly, though it is true that objectives are important at all levels, they are particularly relevant to projects. These are launched to create the deliverables and capabilities which deliver value to and through the business. Setting and achieving objectives are at the heart of project management, and they are the focus of most of the activities at this level. As a result, risk management has a particular importance for management of projects, as we will explore in Chapter 2.

All Risks Are Uncertain, but Not All Uncertainties Are Risks

This chapter has explored the range of uncertainties facing us in the modern world, which are both extensive and pervasive. However not all uncertainties matter equally, and some do not matter at all. It is important and necessary for us to be able to separate out those uncertainties which matter, and to develop appropriate responses to them. The degree to which something matters can be described in terms of its effect on our ability to achieve our objectives, at whatever level those exist. A pragmatic approach to 'risk' treats it as that subset of uncertainty which matters because if it occurs it will affect achievement of objectives. Within this framework, risk management offers a solution to our need to address uncertainty, since it provides us with a structured way of identifying and managing those sources of potential variation which matter. The definition of risk as 'uncertainty that, if it occurs, will affect achievement of objectives' includes both negative and positive risks, threats and opportunities, both of which are types of uncertainty, differing only in the nature of their impact on objectives. All of this is important for projects, which are designed to achieve specific objectives in order to deliver value and benefits to the organisation.

Chapter 2

Risk and Projects

What's Wrong With Projects?

We have seen in Chapter 1 that the world is uncertain, and that some of those uncertainties pose risks, depending on whether they matter in terms of affecting our ability to achieve our objectives. And it is this link with objectives that makes risk particularly relevant to projects, since projects are intimately associated with objectives. This chapter explores why projects are particularly risky, in order to set the context for risk management in projects and to help us to understand why managing risk effectively is essential if we really want our projects to succeed.

Before we examine the reasons behind the close connection between risk and projects, we need to be clear what we mean by the term 'project'. What are projects and why do we do them?

Project management is served by a number of professional bodies and standard-setting organisations, and not surprisingly, these have each developed their own definition of a project. Examples are given in Table 2.1.

These and other definitions make it clear that projects exist for a very clear reason, or at least they should, and this is frequently defined in a business case or project charter. A project is launched in order to create a set of deliverables that implement an aspect of corporate strategy, often through its contribution to a higher-level programme or portfolio. Ultimately a project delivers benefits to the organisation and its stakeholders, but often these

DOI: 10.4324/9781003431954-2

Table 2.1 Definitions of 'Project'

Organisation	Definition of 'project'
Association for Project Management (2019)	A unique, transient endeavour undertaken to bring about change and to achieve planned objectives.
Project Management Institute (2017a)	A temporary endeavour undertaken to create a unique product, service or result.
International Project Management Association (2015)	A unique, temporary, multi-disciplinary and organised endeavour to realise agreed deliverables within predefined requirements and constraints.
European Commission Centre of Excellence in Project Management (CoEPM²) (2018)	A temporary organisational structure which is setup to create a unique product or service (output) within certain constraints.
International Organization for Standardization (2021)	Temporary endeavour to achieve one or more defined objectives.

benefits do not arise immediately or directly as a result of completing the project itself. More often, the execution of a *project* results in *deliverables* that create a *capability* which needs to be operated or used in order to generate the actual *benefits* which correspond to satisfying an element of *corporate strategy* that was defined in the *business case* for the project. This is illustrated in Figure 2.1.

It appears that there is reasonable consensus on what projects are and why we do them. Indeed, mankind has been performing projects for many thousands of years, though not always labelling them as such. Construction of major enterprises in antiquity such as the Seven Wonders of the ancient world (the Great Pyramid of Giza, the Hanging Gardens of Babylon, the statue of Zeus at Olympia, the temple of Artemis at Ephesus, the mausoleum of Mausolus at Halicarnassus, the Colossus of Rhodes, the Lighthouse of Alexandria) were all projects.

With such a long history of executing projects, one would expect that we would be very successful at it by now. Unfortunately, the data suggest otherwise. The best long-term data on project success come from The Standish Group, whose CHAOS Report analysed project performance from 1994 until recently. Projects were categorised as 'successful', 'failed' or 'challenged'. Initially, this was assessed against three criteria: On Time, On Budget, On

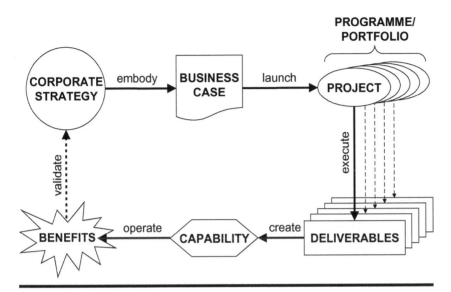

Figure 2.1 Linking Projects to Strategy

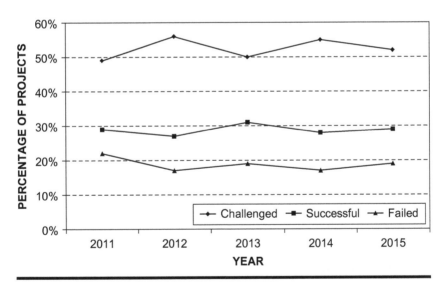

Figure 2.2 Standish CHAOS Data on Project Success 2011–2015 (Data source: The Standish Group, 2015)

Target. However, these were extended in 2015 to six measures, adding On Goal, Valuable and Satisfactory, to reflect the goals of the customer or user of the project deliverables. The CHAOS Report 2015 (The Standish Group, 2015) reassessed project data from 2011 to 2015 against the new criteria, and this is shown in Figure 2.2, indicating that the situation has not improved dramatically over the years. Indeed, when 2011–2015 project data is catego-rised using the original three success measures, more recent outcomes are strikingly similar to historical data from 1994 to 2009.

So why do so many projects fail? It is not due to lack of project man-agement theory, tools and techniques, or trained people. We have a good understanding of project concepts, project management processes are well developed, and the people working on projects are mostly professional, com-mitted and capable. It seems that one of the major reasons for project failure is the occurrence of unforeseen events which disrupt the smooth running of the project and cause irrecoverable deviation from the plan – 'Events, dear boy, events', as former British Prime Minister Harold Macmillan put it when asked by a journalist what was most likely to throw a government off course.

On any given project, some of these unforeseen events were probably unforeseeable. But others are likely to have been knowable, if only someone on the project team had looked in the right place or been aware of what lay ahead. These knowable uncertainties fall under the heading of risks, as future events that, if they occurred, would affect achievement of project objectives.

Why Are Projects Risky?

There seems little doubt that projects are risky, as anyone who has ever worked on one will know. In fact, there are three distinct and separate reasons for this, which we need to understand if we are to manage risk in projects suc-cessfully. These are discussed below under three headings:

1. Common characteristics;
2. Deliberate design;
3. External environment.

Common Characteristics

All projects share a range of features which inevitably introduce uncertainty. Many of these characteristics are described in the definitions of 'project' in Table 2.1. Factors found in all projects which make them inherently risky include:

- *Uniqueness.* Every project involves at least some elements that have not been done before, and naturally there is uncertainty associated with these elements.

- *Complexity.* Projects are complex in a variety of ways, and are more than a simple list of tasks to be performed. There are various kinds of complexity in projects, including technical, commercial, interfaces or relational, each of which brings risk into the project.
- *Assumptions and constraints.* Project scoping involves making a range of guesses about the future, which usually include both assumptions (things we think will or will not happen) and constraints (things we are told to do or not do). Assumptions and constraints may turn out to be wrong, and it is also likely that some will remain hidden or undisclosed, so they are a source of uncertainty in most projects.
- *People.* All projects are performed by people, including project team members and management, clients and customers, suppliers and sub-contractors. All of these individuals and groups are unpredictable to some extent, and introduce uncertainty into the projects on which they work.
- *Stakeholders.* These are a particular group of people who impose requirements, expectations and objectives on the project. Stakeholder requirements can be varying, overlapping and sometimes conflicting, leading to risks in project execution and acceptance.
- *Change.* Every project is a change agent, moving from the known present into an unknown future, with all the uncertainty associated with such movement.

These risky characteristics are built into the nature of all projects and cannot be removed without changing the project. For example, a 'project' which was not unique, had no constraints, involved no people, and did not introduce change would in fact not be a project at all. Trying to remove the risky elements from a project would turn it into something else, but it would not be a project.

Deliberate Design

The definitions of 'project' in Table 2.1 emphasise that projects are conceived, launched and executed in order to achieve objectives which are (or should be) closely linked to corporate strategy. In the competitive business environment, organisations are seeking to get and stay ahead of the competition by making significant advances in the products and services which they offer, and by operating as efficiently and effectively as possible. Many businesses use projects as vehicles to deliver that competitive advantage. Clearly each organisation wishes to move ahead as quickly as possible, and that involves taking risk as the business exposes itself to a range of uncertainties that could affect whether or not it achieves its desired aim. Progress can be made in two ways:

■ One option might be to take small steps, making incremental changes to existing products and services, seeking continuous improvement and evolutionary change. While this strategy might appear to be less risky, it delivers smaller advantages at each increment, and relies on a constant supply of value-enhancing developments.

■ An alternative is to be revolutionary, looking for major innovations and paradigm-breaking change, trying to leapfrog the competition and get several steps ahead. This is a riskier strategy but the potential gains are larger and might be achieved more quickly.

The two strategies reveal an important relationship between risk and reward: they are positively correlated. Higher risk means potentially higher reward, though clearly there is also increased possibility of significant loss. By trying to make bigger changes more quickly, an organisation takes more risk in both dimensions, both positive and negative. This is illustrated graphically in Figure 2.3. For example, attempting to launch a new product in a new market could give first-mover advantage and be very profitable, or it could result in significant losses (shown as position 'A' in Figure 2.3). If on the other hand the organisation plays safe and takes less risk, the potential gains are lower (position 'B').

In project-based organisations, the role of projects is to deliver value-creating capabilities. As a result, projects are deliberately designed as risk-taking ventures. Their specific purpose is to produce maximum reward for the business while managing the associated risk. Since the existence of projects is so closely tied to reward, it is unsurprising that they are also intimately involved with risk. Organisations which understand this connection deliberately design their projects to take risk in order to deliver value. Indeed, projects are undertaken in order to gain benefits while taking the associated risks in a controlled manner.

External Environment

Projects are not conducted in a vacuum, but exist in an environment external to the project itself which poses a range of challenges and constraints. This

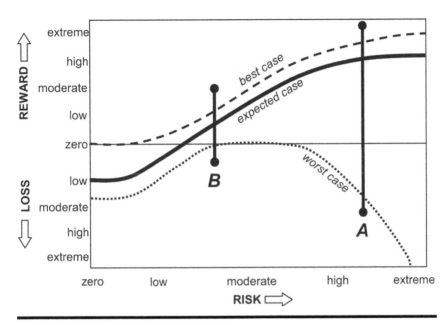

Figure 2.3   Relationship Between Risk and Reward/Loss (Indicative)

includes both the wider organisation beyond the project and the environment outside the organisation, and changes which are outside the project's
control can occur in both of these. Environmental factors which introduce
risk into projects include:

- Market volatility;
- Competitor actions;
- Emergent requirements;
- Client organisational changes;
- Internal organisational changes;
- PESTLIED (political, economic, social, technological, legal, interna-
  tional, environmental, demographic) factors.

Each of these factors is subject to change at an increasing rate in the modern world. Projects essentially have a fixed scope which they are required to
deliver within this ever-changing environment, which naturally poses risk to
the project. It is not possible to isolate most projects from their environment,
so this represents a common source of risk for projects.

Why Manage Risk in Projects?

It is undoubtedly true that projects are risky as a result of their *common characteristics,* by *deliberate design,* and because of the *external environment* within which they are undertaken. It is impossible to imagine a project without risk. Of course, some projects will be high-risk, while others have less risk, but all projects are by definition risky to some extent. The 'zero-risk project' is an oxymoron and a logical impossibility – it does not and cannot exist. But the link between risk and reward makes it clear that not only is a project without risk impossible, it is also undesirable.

The important thing is not to keep risk out of projects, but to ensure that the inevitable risk associated with every project is at a level which is acceptable to the sponsoring organisation, and is effectively managed. Indeed, those involved with launching, sponsoring and managing projects in organisations should welcome risk in their projects, since it enables and supports change, innovation and creativity – as long as it is taken sensibly, intelligently and appropriately, and as long as it is managed effectively. It is also important to remember that not all risk is bad, since the concept includes both threats and opportunities, as discussed in the previous chapter. Within the project context, this means that there are uncertainties that matter because if they occurred, they would hinder achievement of project objectives (threats), but there are also uncertainties whose occurrence would help to achieve those objectives (opportunities).

This of course is why risk management is such an important part of effective project management: since all projects are exposed to risk (both negative threats and positive opportunities), successful projects are the ones where that risk is properly managed.

In outlining the importance of managing risk in projects, we have used words such as 'sensible', 'intelligent', 'appropriate' and 'effective' to describe how risk management should be implemented. The next chapter describes a risk process that embodies those characteristics, but first there is one additional aspect of risk in projects that needs to be clarified.

Individual Risks or Overall Risk (or Both)?

When considering risk in projects, there are two levels of interest, typified by the scope of responsibility and authority of the project manager and the project sponsor:

- The project manager is accountable for delivery of the project objectives, and therefore needs to be aware of any risks that could affect that delivery, either positively or negatively. Their scope of interest is focused on specific sources of uncertainty within the project. These sources are likely to be particular future events or sets of circumstances or conditions which are uncertain to a greater or lesser extent, and which would have some degree of impact on the project if they occurred. The project manager asks, 'What are the risks in my project?', and the answer is usually recorded in a Risk Register or similar document.
- The project sponsor on the other hand is interested in risk at a different level. They are less interested in specific individual risks within the project, and more in the overall picture. Their question is 'How risky is my project?', and the answer does not usually come from a Risk Register. Instead of wanting to know about individual risks, the project sponsor is concerned about the overall riskiness of the project. This represents their exposure to the effects of uncertainty across the project as a whole.

These two different perspectives reveal an important dichotomy in the nature of risk in the context of projects. A project manager is interested in 'individual risks' while their sponsor wants to know about 'overall risk'. While the project manager looks at the risks *in* the project, the project sponsor looks at the risk *of* the project.

This distinction is described in standard approaches to project risk management. Examples where these different levels of risk are defined are provided in Table 2.2, from risk management guidelines published by the Association for Project Management (APM) and the Project Management Institute (PMI) respectively, together with one pair of definitions from an established project risk management methodology (Hillson & Simon, 2020).

Given these two levels of interest, any approach to risk management in projects needs to be able to answer the questions of both project manager and project sponsor. An effective project risk management process should identify individual risk events within the project and enable them to be managed appropriately, and should also provide an indication of overall project risk exposure. This second aspect is less well developed in current thinking and practice, and is the subject of active development by leading practitioners and professional bodies.

Table 2.2   'Risks' vs. 'Risk' in Project Risk Management Guidelines

Organisation	Lower-level 'risks'	Higher-level 'risk'
Association for Project Management (2008)	An *individual risk event* is an uncertain event or set of circumstances that, should it occur, will have an effect on achievement of one of more project objectives.	*Overall project risk* results from an accumulation of a number of individual risk events, together with other sources of uncertainty to the project as a whole, such as variability and ambiguity.
Association for Project Management (2024)	A *risk event* is an uncertain event or set of circumstances that, should it or they occur, will have an effect on achievement of one of more of the project's objectives.	*Project risk* is the exposure of stakeholders to the consequences of variations in outcome.
Project Management Institute (2019)	An *individual risk* is an uncertain event or condition that, if it occurs, has a positive or negative effect on one of more objectives.	*Overall risk* is the effect of uncertainty on the portfolio, program, or project as a whole.
ATOM Methodology (Hillson & Simon, 2020)	A *risk event* is an uncertain discrete occurrence that, if it occurs, would have a positive or negative effect on achievement of one or more objectives.	*Overall project risk* is the exposure of project stakeholders to the consequences of variation in project outcomes. Overall project risk is more than the sum of individual risk events, and includes the effects of other sources of uncertainty such as ambiguity and variability.

Why Is Risk Management Important to Projects?

This chapter has described why projects are risky: by nature, by design and by context. In a real sense, the whole discipline of project management can be seen as an attempt to bring structure and order to the various elements of uncertainty within a project. For example, the purpose of the Work Breakdown Structure (WBS) is to define the full scope of the project, to ensure that this is clearly stated and understood, and to form a basis for project control and monitoring. With a properly defined WBS, there should be no uncertainty about project scope – all project work is described in the WBS, and if it is not in the WBS it is not in the project. Similarly, the Organisational Breakdown Structure (OBS) and Cost Breakdown Structure (CBS) seek to define the roles within the project and the structure of the project budget respectively, in order to reduce or remove possible ambiguity, confusion or misunderstanding. The project schedule describes the dependency relationships between project activities and their expected time-phasing, to reduce uncertainty about 'what happens when'.

While each of the project management disciplines can be seen as addressing some aspect of project uncertainty, it is risk management which has the most direct relevance here, since it specifically and intentionally focuses on those uncertainties that matter. The whole purpose of the risk process is to identify risks and enable them to be managed effectively. As a result, risk management is essential for project success. The outcome of managing risks properly on a project is to reduce the number of threats that materialise into problems, and to minimise the effect of those which do occur. It also results in more opportunities being captured proactively and turned into positive benefits for the project. Effective risk management minimises threats, maximises opportunities and optimises the achievement of project objectives. The converse is also true (as illustrated by the experience of many projects where risk management is less than fully effective). Failing to manage risks on projects will result in more problems, fewer benefits and a lower chance of project success. In this sense, risk management is a true 'CSF' for projects: it is unlikely that projects will be successful without effective management of risk (it is a 'Critical Source of Failure'), and where risk management is working properly projects have the best chance of succeeding (it is a 'Critical Success Factor'). This duality of the importance of effective management of risk for project success is illustrated in Figure 2.4.

Having explained why risk management matters to projects, the next question is how to do it, which is addressed in the next chapter.

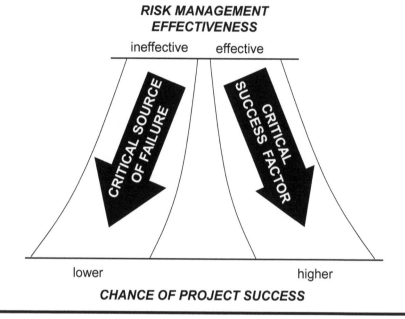

Figure 2.4   Risk Management as a CSF for Project Success

Chapter 3

Managing Risk in Practice

We have seen in the previous chapter that all projects are risky. This arises from their *common characteristics,* as unique and complex undertakings based on assumptions and constraints, delivering change to multiple stakeholders with different requirements. Risk is also a factor in the *deliberate design* of projects, which are launched in order to take sensible levels of risk and thereby gain appropriate rewards for the sponsoring organisation. Finally, projects are risky because of the *external environment* in which they operate, which is characterised by change in many different aspects, all of which create challenges to project success.

While many of the disciplines of project management can be seen as an attempt to address some elements of risk in projects, it is clearly the specific role of risk management to allow both overall project risk and individual risks to be understood, assessed and managed. To do this in an effective way requires a structured process. However, structure can hinder effectiveness if it imposes a bureaucratic or counterintuitive straitjacket on project team members. In order to support the right behaviour and produce the desired outcome, a structured process should reflect the natural way in which people think and act. Fortunately, the typical risk management process meets this requirement, since it simply embodies the way people consider and respond to uncertainty. This chapter starts from first principles and describes a natural approach to dealing with risky situations. The informal principles are then

DOI: 10.4324/9781003431954-3

developed into a structured generic risk process which can be widely applied in a variety of situations, including the management of risk on projects. The chapter addresses how to identify, assess and manage both individual risks and the overall riskiness of the project.

Towards a Risk Management Process

Anyone undertaking a risky or important venture is likely to ask themselves a series of simple questions, namely:

- What are we trying to achieve?
- What could affect us achieving this?
- Which of those things are most important?
- What shall we do about them?
- Who needs to know about them?
- Having taken action, what has changed?
- What did we learn?

These questions represent the simplest expression of an intuitive risk management process. They can be expanded into a more detailed narrative description of a process which corresponds to a natural and logical approach for managing risk in a project context, and this section presents such an expansion. The link between this natural narrative and a formal risk management process can then be made, indicating the extent to which risk management is simply structured common sense.

Getting Started

The definitions of risk shown in Table 1.2 make it clear that risks only exist in relation to defined objectives. This means we cannot start the risk process without first clearly defining its scope, in other words clarifying which objectives are at risk. It is also important to know how much risk key stakeholders are prepared to accept in the project, since this provides the target threshold for risk exposure on the project. These factors must be addressed in the first step of any risk process, to ensure that scope and objectives are well defined and understood.

Finding Risks

Once the scope and objectives are agreed, it is possible for us to start identifying individual risks, which are those uncertainties with the potential to affect achievement of one or more of our objectives (including both threats and opportunities). We could use a variety of techniques, each of which has strengths and weaknesses, so it would be wise to use more than one approach to ensure that as many risks as possible are identified. The aim is to expose and document all currently knowable individual risks, recognising that some risks will be inherently unknowable and others will emerge later in the project. This is why the risk process needs to be iterative, coming back later to find risks which were not evident earlier on. In addition to considering individual risks within the project, we should also aim to uncover sources of overall risk exposure in the project.

Setting Priorities

Of course, not all the individual risks we identify are equally important, so we need to filter and prioritise them, to find the worst threats and the best opportunities. This will inform how we respond to risks. When prioritising individual risks, we could use various characteristics, such as how likely they are to happen, what they might do to project objectives, how easily we can influence them, when they might happen, and so on. We should also consider the degree of overall project risk exposure, either by categorising individual risks to find out whether there are any significant hot-spots or concentrations of risk exposure, or by using simulation models to analyse the combined effect of individual risks on the final project outcome.

Deciding What to Do

Once we have prioritised individual risks and understood the degree of overall project risk exposure, we can start to think about what actions are appropriate to deal with individual threats and opportunities, as well as considering how to tackle overall project risk. We might consider radical action such as cancelling the project, or decide to do nothing, or attempt to influence the level of risk exposure. We should also look for someone who can make a difference and involve them in responding appropriately to the risks.

Taking Action

Of course, we can make great plans to address the individual risks in our project, as well as its overall riskiness, but nothing will change unless we actually do something. Planned responses must be implemented in order to tackle individual risks and change the overall risk exposure of the project, and the results of these responses should be monitored to ensure that they are having the desired effect. Our actions may also introduce new risks for us to address.

Telling Others

After completing these various steps, we are in a position where we know what the individual risks are and how they would affect the project, as well as understanding potential sources of overall project risk, and we understand which ones are particularly significant. We have also developed and implemented targeted responses to tackle our risk exposure, with the help of others. It is important to tell people with an interest in the project about the risks we have found and our plans to address them.

Keeping Up to Date

We have clarified our objectives and found the individual risks that could affect them, then prioritised the important ones and developed suitable actions, and we also know what contributed to the overall risk exposure of our project – so have we finished? Actually no, because risk poses a dynamic and changing challenge to our project. As a result, we know that we have to come back and look again at risk on a regular basis, to see whether our planned actions have worked as expected, and to discover new and changed risks that now require our attention.

Capturing Lessons

When the project ends, should we heave a sigh of relief and move quickly on to the next challenge? As responsible professionals we will wish to take advantage of our experience on this project to benefit future projects. This means we will spend time thinking about what worked well and what needs improvement, and recording our conclusions in a way that can be used by ourselves and others.

From Narrative to Reality

The steps outlined above comprise the logical components of the project risk management process, and these correspond to the steps found in various versions of that process as captured in risk management standards and guidelines. The different steps may be given a range of descriptive titles, but the essential process remains constant. For the remainder of this chapter, we will use slightly more formal names for the process steps, as shown in Table 3.1. Figure 3.1 takes those steps and links them in an iterative process which is repeated throughout the life of the project. Table 3.2 maps the generic risk process steps to some of the most widely used risk standards, indicating the degree of commonality.

One interesting and important point arises from the comparison of risk standards and guidelines in Table 3.2, namely that most lack a final step at the closure of the project to learn risk-related lessons for the benefit of future projects and the wider organisation. Among the risk standards listed, only IEC62198:2024 includes any mention of the need to capture lessons, and this is only covered in two sentences as part of a wider 'Monitoring and Review' step. This absence in the majority of commonly used standards and guidelines reflects a wider malaise: the reluctance of many organisations to undertake a post-project review or lessons learned exercise at the end of their completed projects (or at significant intermediate milestones). For some reason it seems that the effort to perform such a review is too much for most, despite the obvious benefits that can accrue. Perhaps this is because those benefits come too late to help the completed project and project teams lack the necessary altruism to help those who come after them. Or maybe it is simply a practical matter of staff being allocated to the next job before they have time to capture the lessons that could be derived from their recent experience. Whatever the reason, organisations that fail to conduct post-project reviews are denying themselves the benefit of experience, and are increasing the chances of repeating the same mistakes in future. This applies to the risk process as much as to any other aspect of project management. There are risk-related lessons to be learned from every project, and ideally these should be captured during a routine post-project review exercise. Where such a wider step is missing from the project process, it should at least be included in the risk process, as in the generic risk process described in this chapter.

Table 3.1 Informal and Formal Risk Process Steps

Informal process step	Formal process step	Purpose
Getting started [*What are we trying to achieve?*]	Risk Process Initiation	To define the scope, objectives and practical parameters of the project risk management process.
Finding risks [*What could affect us achieving this?*]	Risk Identification	To identify all currently knowable risks, including both individual risks and sources of overall project risk.
Setting priorities [*Which of those things are most important?*]	Qualitative Risk Assessment	To evaluate key characteristics of individual risks enabling them to be prioritised for further action, and recognising patterns of risk exposure.
	Quantitative Risk Analysis	To evaluate the combined effect of individual risks and other forms of uncertainty on the project outcome and assess overall project risk exposure.
Deciding what to do [*What shall we do about them?*]	Risk Response Planning	To determine appropriate response strategies and associated actions for each individual risk and for overall project risk, together with suitable response/action owners.
Taking action [*Do it!*]	Risk Response Implementation	To implement agreed actions, determine whether they are working, and identify any resultant secondary risks.
Telling others [*Who needs to know about them?*]	Risk Communication	To inform project stakeholders about the current level of risk exposure and its implications for project success, including both individual risks and overall project risk, as appropriate.
Keeping up to date [*Having taken action, what has changed?*]	Risk Review	To review changes in identified risks and overall project risk exposure, identify additional actions as required, and assess the effectiveness of the project risk management process.
Capturing lessons [*What did we learn?*]	Post-Project Risk Review	To identify risk-related lessons to be learned for future projects.

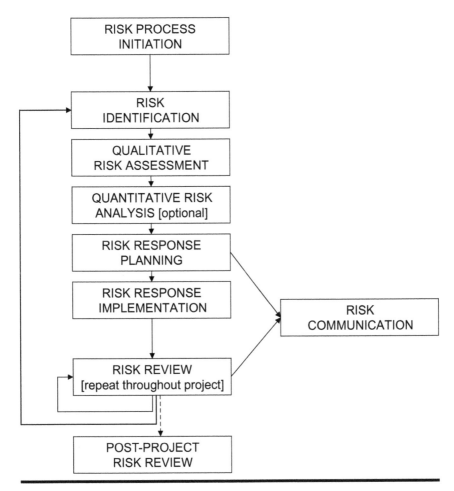

Figure 3.1   Risk Process

Describing the Risk Process

Having developed from first principles a generic process for managing risk on projects, we can now describe what is entailed in each step. The description that follows is necessarily high level and does not present all of the possible tools and techniques in exhaustive detail. Such information is available from a wide variety of risk management textbooks and training courses, and is outside the scope of this book. Instead, we present the key techniques involved at each step with sufficient detail to ensure that their purpose is understood. It is also important to remember that the risk process can be implemented at different levels, from a few simple and informal steps to a

Table 3.2 Mapping Generic Risk Process to Risk Standards

Informal process step	Formal process step	ISO31000:2018 Risk Management – Guidelines (International Organization for Standardization, 2018)	Management of Risk (M_o_R4). (Axelos, 2022)	IEC62198:2024 Managing risk in projects – Application guidelines (International Electrotechnical Commission, 2024)	Project Risk Analysis and Management (PRAM) Guide. (Association for Project Management, 2024)	The Standard for Risk Management in Portfolios, Programs and Projects. (Project Management Institute, 2019)
Getting started	Risk Process Initiation	Scope, Context, Criteria	Define context and objectives	Scope, context and criteria	Initiate	Plan Risk Management
Finding risks	Risk Identification	Risk Identification	Identify threats and opportunities	Risk identification	Identify	Identify Risks
Setting priorities	Qualitative Risk Assessment	Risk Analysis Risk Evaluation	Prioritize risks	Risk analysis, Risk evaluation	Assess	Perform Qualitative Risk Analysis
	Quantitative Risk Analysis		Assess combined risk profile	Quantitative Risk Analysis using simulation		Perform Quantitative Risk Analysis
Deciding what to do	Risk Response Planning	Risk Treatment	Plan responses Agree contingency	Selection of risk treatment options	Plan Responses	Plan Risk Responses
Taking action	Risk Response Implementation	Preparing and implementing risk treatment plans	-	Risk treatment plans	Implement Responses	Implement Risk Responses
Telling others	Risk Communication	Recording and Reporting Communication and Consultation	Monitor and report progress	Communication and consultation Recording and reporting	-	-
Keeping up to date	Risk Review	Monitoring and Review	Review and adapt	Monitoring and review	Manage Process	Monitor Risks
Capturing lessons	Post-Project Risk Review	-	-	Monitoring and review	-	-

fully rigorous and integrated process. (The scalable nature of project risk management is discussed later in this chapter.) Project-based organisations would be well advised to adopt a fully scalable risk process that embodies the good practices outlined here (an example such process is detailed by Hillson & Simon, 2020).

Most of the process described in the following sections is aimed at individual risks, but consideration of overall project risk is also highlighted where appropriate.

Risk Process Initiation

The first step of the risk management process is not Risk Identification. In Chapter 1 we developed a definition of risk as 'uncertainty that, if it occurs, will affect achievement of objectives'. Since risk is defined in terms of objectives, the essential first step of the risk process is to define those objectives which are at risk. This gives us the scope of the risk process, and is the main purpose of the Risk Process Initiation step.

It is also important to recognise that risk management is not 'one-size-fits-all'. Since every project has a different level of risk exposure, it is necessary to scale the risk process to meet the risk challenge of each particular project. Projects which are highly risky or strategically important will require a more robust approach to risk management than those which are simpler or routine. Scalable aspects of the risk process are discussed in more detail later in this chapter, and include organisation and staffing, methodology, tools and techniques, infrastructure, reporting requirements, and the review and update cycle. The depth and complexity of the risk process which is to be applied to the particular project at hand needs to be decided and documented during the Risk Process Initiation step.

This step also involves a number of other important decisions which must be made before we can start the risk management process. The first of these is to set thresholds for how much risk is acceptable on this particular project, stated against each of the key project objectives. Examples of risk threshold statements might include the following:

■ *Schedule.* There is no flexibility in the final project delivery date, which is required to meet a fixed client launch date. If early delivery is predicted, the project manager will discuss possible product enhancements with the project sponsor.

- *Budget.* Use of the allocated project contingency fund is acceptable, but any further cost overrun up to 5 per cent of budget must be authorised in advance by the project sponsor. Projected overspend of >5 per cent will trigger a strategic review and possible project termination. Projected underspend of up to 10 per cent is permitted, but additional options for cost savings must be notified to the project sponsor to allow possible budget reallocation to other projects.
- *Performance.* No performance variation is permitted in features identified as 'critical' in the design documentation. Performance of 'secondary' features may vary by +/− 10 per cent. Predicted variations outside this limit must be notified immediately to the System Architect and may result in design modifications.

In order to define the risk thresholds for the project, we first need to identify the risk tolerances of key stakeholders. Extracting risk tolerances from stakeholders can be difficult, since these individuals are often not explicitly aware of how much risk they are prepared to take. In addition, it is likely that different stakeholders will have different tolerances to risk, and this will require discussion to reach consensus on what risk thresholds should be applied. While the project sponsor should take the lead in these discussions as part of their responsibility to develop the business case for the project, it is often the case that the project manager will be closely involved in this step.

When agreement has been obtained on appropriate risk thresholds, it is then possible to transform these into definitions of the scales to be used for qualitative assessment of probability and impact on the project, related to specific project objectives. It is common to use terms such as 'high, medium, low' for this purpose, and their meanings must be agreed in advance in order to provide a consistent framework for assessment of identified individual risks. A definitions table similar to the example in Table 3.3 should be produced, which reflects the agreed risk thresholds for this project. This provides a single common framework which can be used to assess individual risks across the project.

A final component of the Risk Process Initiation step is to define potential sources of risk to the project. This is often presented as a hierarchical Risk Breakdown Structure (RBS), perhaps drawing on an industry standard or an organisational template (Hillson, 2003). An example RBS is given in Table 3.4. The RBS can be used as a framework for risk identification and assessment, and to structure the post-project risk review. It can be used both for individual risks, and also to explore potential sources of overall project risk exposure.

Table 3.3   Defining Terms for Probability and Impacts to Reflect Project Risk Thresholds

Scale	Probability	+/- Impact on project objectives		
		Time	Cost	Performance
VHI	76-95%	>20 days	>$100K	Very significant impact on overall functionality
HI	61-75%	11-20 days	$51K-$100K	Significant impact on overall functionality
MED	41-60%	4-10 days	$11K-$50K	Some impact in key functional areas
LO	26-40%	1-3 days	$1K-$10K	Minor impact on overall functionality
VLO	5-25%	<1 day	<$1K	Minor impact on secondary functions
NIL	<5%	No change	No change	No change in functionality

Note: When using these impact scales to assess opportunities, they are to be treated as representing a positive saving in time or cost, or increased functionality. For threats, each impact scale is interpreted negatively, that is, time delays, increased cost or reduced functionality.

A number of important scoping decisions are made during this Risk Process Initiation step, and these need to be documented and communicated to the project team and other key stakeholders. The key output from this step is a clear definition of the scope of the risk process to be employed for this particular project, and this is documented in a Risk Management Plan. The plan should be reviewed from time to time during the project, and must be updated if the risk process is modified. A sample contents list for a typical Risk Management Plan is given in Table 3.5.

Risk Identification

Since it is not possible to manage a risk which has not first been identified, some view Risk Identification as the most important step in the risk process. There are many good techniques available for risk identification, the most common of which include:

■ Use of brainstorming in a facilitated workshop setting, perhaps structured into a SWOT Analysis to identify organisational strengths/weaknesses and project opportunities/threats.
■ Checklists or prompt lists to capture learning from previous risk assessments.

Table 3.4 Example Risk Breakdown Structure (RBS)

RBS level 0	RBS level 1	RBS level 2
0. All risks	1. Technical risk	1.1 Scope definition
		1.2 Requirements definition
		1.3 Technical processes
		1.4 Technology
		1.5 Technical interfaces
		1.6 Design
		1.7 Performance
		1.8 Reliability & maintainability
		1.9 Safety & security
		1.10 Test & acceptance
	2. Management risk	2.1 Project management
		2.2 Programme/portfolio management
		2.3 Operations management
		2.4 Organisation
		2.5 Resourcing
		2.6 Communication
		2.7 Information
		2.8 Health, Safety & Environment
		2.9 Quality
		2.10 Reputation
	3. Commercial risk	3.1 Contractual terms & conditions
		3.2 Warranties & liabilities
		3.3 Internal procurement
		3.4 Suppliers & vendors
		3.5 Subcontracts
		3.6 Client/customer stability
		3.7 Partnerships & joint ventures
		3.8 Force majeure
		3.9 Dispute resolution/arbitration
		3.10 Intellectual Property

(Continued)

Table 3.4 (Continued)   Example Risk Breakdown Structure (RBS)

		4.1	Legislation
		4.2	Exchange rates
		4.3	Site/facilities
		4.4	Environmental/weather
	4. External risk	4.5	Competition
		4.6	Regulatory
		4.7	Political/Country
		4.8	Country
		4.9	Social/demographic
		4.10	Pressure groups

Table 3.5   Risk Management Plan Sample Contents List

Introduction
Project description and objectives
Aims, scope and objectives of risk process
Risk tools and techniques
Organisation, roles and responsibilities for risk management
Risk reviews and reporting
Appendices
A Project-specific definitions of probability and impacts
B Project-specific sources of risk (Risk Breakdown Structure)

- Detailed analysis of project assumptions and constraints to expose those which are most risky.
- Interviews with key project stakeholders to gain their perspective on possible risks facing the project.
- Review of completed similar projects to identify common risks and effective responses.

For each of these techniques, it is important to involve the right people with the necessary perspective and experience to identify risks facing the project. It is also helpful to use a combination of risk identification techniques rather than rely on just one approach – for example, perhaps using a creative group technique such as brainstorming together with a checklist based on past similar projects. The project manager should select appropriate techniques based on the risk challenge faced by the project, as defined in the Risk Management Plan.

The project's RBS can be used as a framework for risk identification, to make sure that all possible sources of risk are considered, to identify gaps and to act as a prompt list. The same RBS can be used to support identification of both individual risks and sources of overall project risk. Alternatively, higher-level prompt lists can guide thinking around what might cause the project as a whole to be more or less exposed to risk; common structures include:

- PESTLE = Political, Economic, Social, Technological, Legal, Environmental;
- PESTLIED = PESTLE + International (or Informational) and Demographic;
- STEEPLE = PESTLE + Ethics;
- InSPECT = Innovation, Social, Political, Economic, Communications, Technology;
- SPECTRUM = Socio-cultural, Political, Economic, Competitive, Technology, Regulatory/legal, Uncertainty/risk, Market;
- TECOP = Technical, Environmental, Commercial, Operational, Political.

It is also a good idea to look out for immediate 'candidate' responses during the Risk Identification phase. Sometimes an appropriate response becomes clear as soon as the risk is identified, and in such cases, it might be advisable to tackle the risk immediately, if possible, as long as the proposed response is cost-effective and feasible.

Whichever technique is used, it is important to remember that the aim of Risk Identification is to identify individual risks and sources of overall project risk. While this may sound self-evident, in fact this step in the risk management process often exposes things which are not risks, including problems, issues or complaints. The most common mistake is to identify either causes of risks (which are present conditions that give rise to risks) or the effects of risks (the direct impact that a risk would have on an objective if it happened), and to confuse these with individual risks. Including causes or effects in the list of identified risks can obscure genuine risks, which may not then receive the appropriate degree of attention they deserve. One way to clearly separate risks from their causes and effects is to use *risk metalanguage* to provide a three-part structured 'risk statement', as follows:

As a result of *<definite cause>,*
    <uncertain event> may occur,
        which would lead to *<effect on objective(s)>.*

Examples of good risk statements include the following:

- As a result of using novel hardware (a definite requirement = cause), unexpected system integration errors may occur (an uncertain event = risk), which would lead to overspend on the project (an effect on the budget objective).
- Because our organisation has never done a project like this before (fact = cause), we might misunderstand the customer's requirement (uncertainty = risk), in which case our solution would not meet the performance criteria (contingent possibility = effect on objective).
- We have to outsource production (cause), so we may be able to learn new practices from our selected partner (risk), leading to increased productivity and profitability (effect).

The use of risk metalanguage should ensure that Risk Identification actually identifies individual risks, distinct from causes or effects. Without this discipline, Risk Identification can produce a mixed list containing risks and non-risks, leading to confusion and distraction later in the risk process.

Many of the most common risk identification techniques focus on individual risks within the project, but are not generally used to consider sources of overall risk to the project. These are also important and should be identified in a structured way. They include uncertainties around the scope and purpose of the project, its role in delivering benefits to the wider organisation, and other types of ambiguity where available information is insufficient.

Each identified individual risk or source of overall project risk should be allocated to a risk owner who will be responsible for ensuring that it is managed effectively.

Having used a variety of techniques to find risks, the Risk Identification step ends by ensuring that individual risks are documented in the Risk Register. The format of Risk Registers can be simple or complex, depending on the information requirements of the project and the sponsoring organisation, and this is one of the scalable aspects of the risk process defined in the Risk Management Plan. Where software tools are used to support the risk process, these usually offer a Risk Register format, though some

Table 3.6   Typical Risk Register Data

Project data	Project Reference Number, Project title Project Manager Client
Risk data	Unique risk identifier Risk type (threat or opportunity) RBS reference (source of risk) WBS reference (area affected by risk) Risk title Risk description (cause-risk-effect) Risk status Risk owner Date risk raised
Assessment data	Probability of occurrence – rating Impacts against objectives – rating & description Related risks
Response data	Preferred response strategy Actions to implement strategy Action owners Action planned start and completion dates Action status Secondary risks Trigger conditions Review date Date risk closed/deleted/expired/occurred

organisations develop their own. The Risk Register is updated following each of the subsequent steps in the risk process, to capture and communicate risk information and allow appropriate analysis and action to be undertaken. The type of data held in a typical Risk Register is listed in Table 3.6.

Qualitative Risk Assessment

Risk Identification usually produces a long list of individual risks, perhaps categorised in various ways. However, it is not usually possible to address all risks with the same degree of intensity, due to limitations of time and resources. And not all risks deserve the same level of attention. It is therefore necessary to be able to prioritise individual risks for further consideration, in order to identify the worst threats and best opportunities. This is the purpose of Qualitative Risk Assessment.

The definition of risk as 'uncertainty that, if it occurs, will affect achievement of objectives' indicates that risk has at least two important dimensions: uncertainty, and its potential effect on objectives. The term 'probability' is usually used to describe the uncertainty dimension, though other terms such as 'frequency' or 'likelihood' are also common. 'Impact' is most often used to describe effect on objectives, or sometimes 'consequence'. For qualitative assessment, these two dimensions are assessed using labels such as 'high, medium, low', which have been previously defined in the Risk Management Plan (see the earlier example in Table 3.3). The probability of each individual risk occurring is assessed, as well as its potential impact if it were to occur. Impact is assessed against each project objective, usually including time and cost, and possibly others such as performance, quality, regulatory compliance and so on. For threats, impacts are negative (lost time, extra cost and so on), but opportunities have positive impacts (saved time or cost and so on). This assessment is often done by the project team in a workshop setting, although it is possible for the relevant risk owner to assess their own risks.

The two-dimensional assessment is used to plot each risk onto a Probability-Impact Matrix, with high/medium/low priority zones. These zones are often coloured following a traffic-light convention, with red used for high-priority risks to be treated urgently, amber/yellow designating risks of medium priority to be monitored, and the green zone containing low-priority risks. It is increasingly common to use a double 'mirror' matrix format plotting threats and opportunities separately, and creating a central zone of focus, as shown in Figure 3.2. This zone contains the worst threats (with high probability so they are likely to happen unless managed, and high impact so they would be very bad for the project) and the best opportunities (where high probability means easy to capture, and high impact means very good). Some larger projects may enhance the Probability-Impact Matrix by using a probability-impact scoring scheme similar to the example shown in Figure 3.3. These calculated Probability-Impact (P-I) Scores allow risks to be prioritised in more detail than the simple three-zone traffic-light approach.

Of course, risks have other characteristics in addition to probability and impact, and these can also be assessed and used to prioritise risks for further attention (Association for Project Management, 2008). Such factors might include:

■ The degree to which a risk can be managed (manageability);
■ Its potential to affect the wider organisation directly (relatedness/connectivity);
■ How soon the risk might occur (proximity);
■ The time window when action might be possible (urgency).

The traditional Probability-Impact Matrix does not allow these additional factors to be used in risk prioritisation, and other techniques are required

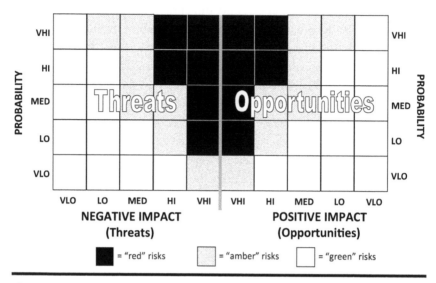

Figure 3.2   Double Probability-Impact Matrix

THREATS   OPPORTUNITIES

PROBABILITY											PROBABILITY
VHI 0.90	0.05	0.09	0.18	0.36	0.72	0.72	0.36	0.18	0.09	0.05	**VHI** 0.90
HI 0.70	0.04	0.07	0.14	0.28	0.56	0.56	0.28	0.14	0.07	0.04	**HI** 0.70
MED 0.50	0.03	0.05	0.10	0.20	0.40	0.40	0.20	0.10	0.05	0.03	**MED** 0.50
LO 0.30	0.02	0.03	0.06	0.12	0.24	0.24	0.12	0.06	0.03	0.02	**LO** 0.30
VLO 0.10	0.01	0.01	0.02	0.04	0.08	0.08	0.04	0.02	0.01	0.01	**VLO** 0.10
	VLO 0.05	**LO** 0.10	**MED** 0.20	**HI** 0.40	**VHI** 0.80	**VHI** 0.80	**HI** 0.40	**MED** 0.20	**LO** 0.10	**VLO** 0.05	
	NEGATIVE IMPACT					**POSITIVE IMPACT**					

RANK	PROB	IMPACT
VHI	0.9	0.8
HI	0.7	0.4
MED	0.5	0.2
LO	0.3	0.1
VLO	0.1	0.05

Figure 3.3   Example Probability-Impact Scoring Scheme

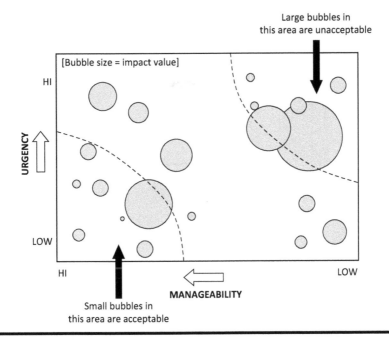

Figure 3.4 Example Bubble Diagram (From Association for Project Management, 2008)

if they are to be taken into account. Common formats for prioritising risks using more than two parameters are the Bubble Diagram and the Risk Prioritisation Chart (examples are shown in Figures 3.4 and 3.5).

The results of the Qualitative Risk Assessment step for each individual risk are documented in the Risk Register, together with any supporting information to justify or explain the basis for the assessment.

Another important output from qualitative assessment is to understand the pattern of risk on the project, and whether there are common causes of risk or hot-spots of exposure. This can be assessed by mapping individual risks into the RBS to determine whether any particular causes are prevalent, and by mapping risks into the Work Breakdown Structure (WBS) to identify areas of the project that might be most affected. This mapping can be conducted simply by counting the numbers of risks in each category, or more accurately by weighting the risks in each category using their P-I Scores.

The techniques mentioned above are useful for prioritising individual risks, but cannot be applied to assess the overall level of project risk exposure. This requires the use of Quantitative Risk Analysis techniques, which are discussed in the following section.

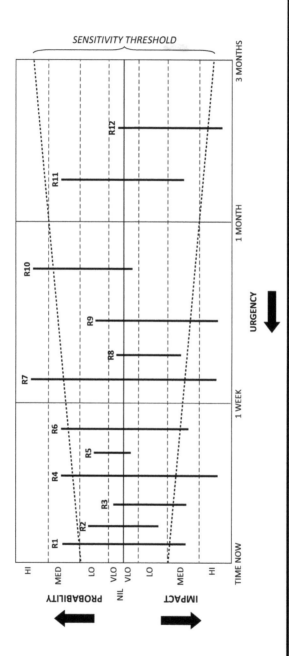

Figure 3.5   Example Risk Prioritisation Chart (Adapted From Barber, 2003; Source: Association for Project Management, 2008)

Quantitative Risk Analysis

On most projects, individual risks do not happen one at a time. Instead, they interact in groups, with some risks causing others to be more likely and some risks making others impossible. For the most part, Qualitative Risk Assessment considers risks individually, and allows development of a good understanding of each one (although grouping risks into categories can give some insights into patterns of risk exposure). It can however sometimes be useful to analyse the combined effect of risks on project outcomes, particularly in terms of how they might affect overall time and cost. Indeed, this is often the only way to obtain an accurate assessment of the overall risk exposure of the project. This is the purpose of the Quantitative Risk Analysis step.

Various Quantitative Risk Analysis techniques are available, including Monte Carlo simulation and decision trees. Monte Carlo is the most popular because it uses simple statistics, it often uses existing project data as its baseline, and there are many good software tools to support it (Hulett, 2009, 2011). Decision trees are however particularly useful for analysing key strategic decisions or major option points.

One key aspect of Quantitative Risk Analysis models which is often overlooked is the need to include both threats and opportunities. If only threats are considered then the analysis is only modelling potential downside, and the result will always be pessimistic. Since the risk process aims to tackle both threats and opportunities, both must be included in any analysis of the effect of risk on the project. Indeed, some vital elements of the risk model such as three-point estimates cannot be properly determined without considering both upside as well as downside. Opportunities relating to planned activities must be taken into account when producing the minimum/ optimistic/best-case estimate, and threats are relevant when estimating maximum/pessimistic/worst-case values.

When developing Monte Carlo risk models, it is easy to use available software tools to create simple (or simplistic) models which do not reflect the complexities of the risks facing the project. In particular, simply taking single values of duration or cost in a project plan or cost estimate and replacing them with three-point estimates is not sufficient to model risk quantitatively. Other modelling techniques should be used to reflect reality, including:

- Different input data distributions, not just the typical three-point estimate (for example, the modified triangular, uniform, spike/discrete, or various curves);

■ Use of stochastic branches to model alternative logic (these can also be used to model key risks);
■ Correlation (also called dependency) between various elements of the model, to ensure that related areas of uncertainty are properly represented.

It is important to recognise that additional investment is required in order to implement Quantitative Risk Analysis, including purchase of software tools, associated training, and the time and effort required to generate input data, run the model and interpret the outputs. Indeed, a particular project may lack the expertise required for conducting a Quantitative Risk Analysis, and may need to bring in help from outside the project. As a result, in many cases the use of quantitative techniques may not always be justified in order to support effective management of risks within the project. Often enough information can be obtained from qualitative assessment, and quantitative analysis techniques can be seen as optional. Many organisations only use Quantitative Risk Analysis for projects which are particularly complex or risky, or where quantitative decisions must be made, for example, concerning bid price, contingency, milestones, delivery dates and so on. However, we need to remember that Quantitative Risk Analysis is the main means of assessing overall project risk exposure, and if it is not used on a particular project then the project stakeholders will be deprived of the wider insights that are only available from this type of approach. It is also possible to implement Quantitative Risk Analysis at different levels of complexity, and often a simple analysis is all that is required in order to give a view of overall project risk. There are several potential shortfalls when using Quantitative Risk Analysis techniques, including:

1 *Poor model quality.* Building a Quantitative Risk Analysis model is not trivial, and requires thought, time and expertise. The level of detail must be right, reflecting key relationships and risks without including unnecessary elements. It is particularly important to include appropriate correlation to reflect linked areas of uncertainty.
2 *Data quality.* It is essential to avoid the GIGO situation (garbage in – garbage out), and attention must be paid to ensuring good quality inputs to the model.
3 *Interpretation.* Outputs from risk models require interpretation, and Quantitative Risk Analysis will not tell the project manager what decision to make.

4 *Action.* The project team must be prepared to use the results of risk modelling, and to take decisions based on the analysis. We should beware of 'analysis paralysis', since Quantitative Risk Analysis is merely a means to an end, and must lead to action.

The main output from a Monte Carlo simulation is the S-curve, presenting a cumulative probability distribution of the range of possible values for the parameter being analysed (for example, total project cost, overall duration, end date and so on). An example is shown in Figure 3.6. Various elements of useful information can be obtained from the S-curve, including:

■ The likelihood of the project meeting its objectives (taken as the cumulative probability of achieving a given target value);
■ The degree of overall uncertainty in the project parameter (derived from the range of possible simulation outputs);
■ The predicted 'expected value' which would occur on balance if the situation remained unmanaged (taken from the mean of all possible results);
■ Output values corresponding to particular confidence levels (for example, the 85th percentile from the S-curve represents the value for which we can have 85 per cent confidence of it not being exceeded).

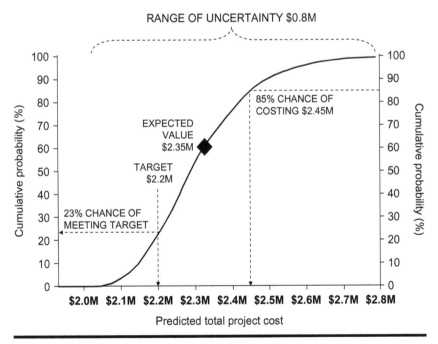

Figure 3.6   Example S-Curve From Monte Carlo Analysis

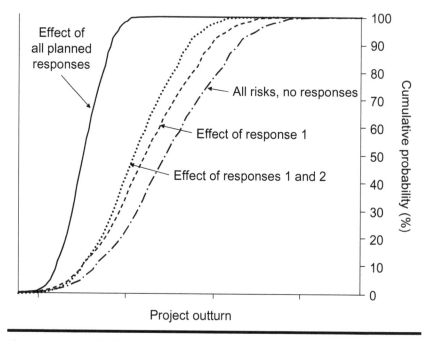

Figure 3.7   Overlaid S-curves

S-curves can be produced for the overall project, or for interim milestones or specific subprojects, allowing analysis of the components of overall project risk. It is also possible to produce overlaid S-curves, as shown in Figure 3.7, indicating the cumulative effect of addressing individual risks, showing the relative contribution of planned responses towards overall project risk exposure.

Risk Response Planning

Having identified and analysed individual risks and the level of overall project risk exposure, it is essential that something should be done in response. As a result, many believe that the Risk Response Planning phase is the most important step in the risk process, since this is where the project team get a chance to make a difference to the risk exposure facing the project. It is usually the responsibility of each risk owner to decide what type of response is most appropriate, though they will often seek help and advice on this.

When developing risk responses, it is important to adopt a *strategic* approach in order to focus attention on what is being attempted. Too often, project teams resort to a 'scatter-gun' approach, trying a wide range of different responses to a given risk, some of which may be counterproductive.

It is better first to select an appropriate strategy for a particular risk, then to design action to implement that strategy, producing a more focused 'rifle-shot' aimed at managing the risk effectively.

Since our definition of risk includes both opportunities and threats, we need to have strategies to deal with both types of risk. Eight possible risk response strategies are available, with a first option for non-project risks, three pairs of proactive options (each pair containing one strategy for threats and a corresponding one for opportunities), and a final last-resort strategy that can be applied to both threats and opportunities, as follows:

- *Escalate.* This is used for threats or opportunities identified by the project team during the project risk management process, but which do not affect project objectives. In these cases, details of the risk are passed to the owner of the objectives that would be affected if the risk occurs. This might be another project, a programme or portfolio, or some other part of the business. Once the risk has been escalated (and acknowledged by the new owner), it does not need to be recorded by the project.
- *Avoid/Exploit.* The goal of these strategies is to remove uncertainty, if possible. For threats the aim of avoidance is to eliminate the risk to the project, making the threat impossible or irrelevant. To exploit an opportunity means to make it definitely happen, ensuring that the project gains the additional benefits.
- *Transfer/Share.* These strategies require involving another person or party in managing the risk. For threats the pain is transferred, together with the responsibility for managing the potential downside. In a similar way the potential gain from an upside risk can be shared, in return for the other party taking responsibility for managing the opportunity.
- *Reduce/Enhance.* Reduction of a threat aims to reduce its probability and/or impact, while enhancing an opportunity seeks to increase them.
- *Accept.* For residual threats and opportunities where proactive action is either not possible or not cost-effective, acceptance is the last resort, taking the risk either without special action or with contingency.

These strategy types are usually only considered as relating to individual risks within a project, but they are also applicable to address the level of overall project risk (with the exception of Escalate). For example:

- The *Avoid* strategy might lead to project cancellation if the overall level of risk remains unacceptable. An *Exploit* response may lead to an agreed expansion of the project scope, or the launch of a new project or subproject.

■ *Transfer/Share* strategies could result in setting up a collaborative business structure in which the customer and the supplier share the risk.

■ *Reduce/Enhance* strategies can be achieved at the overall project level by replanning the project or changing its scope and boundaries.

■ *Accepting* an overall level of project risk means that the project will be continued without significant change, though the organisation may make contingency plans and monitor exposure against predefined trigger conditions.

When choosing a response strategy for an individual risk, the factors used to prioritise risks should be considered again, so that the level of response matches the importance of the risk. For example, the most aggressive response strategies (Avoid/Exploit) should be applied to the highest priority risks if possible, and only the lowest priority risks should be accepted. Unfortunately, response selection is not usually so straightforward, and there are many factors to bear in mind.

It is common for prioritisation to be based only on probability and impact, and so these also typically drive response selection. However, there are other significant risk prioritisation factors (as discussed above) including manageability, relatedness, proximity and urgency. In addition to these, there are also other important considerations which relate specifically to response selection, such as:

■ Availability of resources to address the risk (resourcing);
■ Likely cost of addressing the risk compared to its possible impact (cost-effectiveness);
■ Degree to which the probability and/or impact might be modified (risk-effectiveness);
■ Whether the response will introduce additional risks (secondary risks).

It is clear that selecting the appropriate response to each risk is not a trivial task, and this requires careful thought. The various options should be analysed in order to pick the one most likely to achieve the desired result in a way that is appropriate, affordable and achievable. This is why it is important to maintain energy and focus during the Risk Response Planning step, as discussed in Chapter 9.

Having chosen separate response strategies for each individual risk, as well as identifying strategies to tackle the overall level of project risk exposure, the risk owner (perhaps with the help of the project team) should then develop specific actions to put these strategies into practice, each with an agreed action owner. The selected response strategy and associated actions are documented in the Risk Register.

Risk Response Implementation

It is at this point that most risk management processes fail. Whichever response strategy is selected, it is vital to go from analysis to action, otherwise nothing changes. Unfortunately. many project teams identify and assess risks, develop response plans and complete a Risk Register, then 'file and forget'. Actions are not implemented and the risk exposure remains the same. The problem often arises because the project risk process used by many organisations does not include an explicit step to implement risk responses. Instead, an assumption is made that actions once agreed will of course be implemented. This is dangerous since the project might proceed on the basis that its risks are being managed effectively, when in reality the risk exposure remains unchanged. Without taking effective action, the project will still be at risk to the same degree. For this reason, it is recommended that risk processes should include an explicit Risk Response Implementation step, to bridge the gap between analysis and action.

The key to making sure that risk responses are implemented is not to allow risk responses to be seen as 'extra work', to be done only when project tasks are complete. Risk responses are genuine project tasks, because they are work which has to be done in order for the project to succeed. They should therefore be treated like any other project task. Each risk response should be fully defined, with a duration, budget, resource requirement, completion criteria and so on. The defined action should be allocated to an action owner who has the necessary ability and availability to complete it. A new task should then be added to the project plan for each agreed risk response, and these should be completed, reviewed and reported on like all other project tasks.

An important part of this Risk Response Implementation step is to monitor the effect of actions after they have been taken. For example, the 'risk-effectiveness' of each proposed response should have been considered during the Risk Response Planning step, as an indication of the change in risk exposure which can be expected as a result of implementing the chosen response.

Having completed the action, the risk owner and/or action owner should assess the actual result to decide whether the risk has been changed in the manner predicted. The status of actions and their results are documented in the Risk Register.

One possible side-effect of taking action to address risks is particularly important. In some cases, actions taken in response to one risk may introduce new risks that previously did not exist. Such risks are called 'secondary risks', not because they are less important, but because their existence is dependent on a prior action being completed. Some secondary risks will have been identified during the Risk Response Planning step, but more may come to light when responses are actually implemented. These new risks should be documented and addressed during the risk process as part of the Risk Review step, which is discussed below.

Risk Communication

This step involves producing risk reports at various levels and for different stakeholders. It is important to communicate the results of the risk process, since the aim is to actively manage the risks, and this is likely to require action by stakeholders outside the immediate project team. Risk reports should form a basis for action, and include clear conclusions ('What we have found') and recommendations ('What should be done'). The outputs of the Risk Communication step must be targeted to the information needs of each recipient group, rather than simply issuing Risk Registers to everyone. Some stakeholders might require only a high-level summary of risk exposure at the overall project level, while others might need details of the individual risks that relate to a particular area of the project. In some cases, a graphical 'risk dashboard' might be suitable, and at other times a written report containing detailed analysis and narrative may be necessary.

Risk Communication should be planned and intentional, delivering accurate risk information in a timely manner, targeted to the specific needs of each stakeholder.

Risk Review

The purpose of this step is to ensure that the planned responses are achieving what was expected, and to develop new responses where necessary. It is also important to determine whether new risks have arisen on the project, and to assess the overall effectiveness of the risk management process. These aims are best achieved through a dedicated risk review meeting, though it is possible on smaller projects to review risk as part of a regular project progress meeting. The Risk Review can result in existing risks being left open (sometimes referred to as an active risk), possibly with revised assessments of probability and/or impact. Alternatively, a risk may be marked as closed if it has been resolved or if it has occurred. An existing risk may also be deleted if it is no longer valid or relevant to the project. The Risk Review can also result in identification of new active risks.

The results of the Risk Review step are documented in an update of the Risk Register, and may also result in a new set of risk reports as defined in the Risk Communication step.

Risk management is a cyclic iterative process, and should never be done just once on a project. Risk exposure changes constantly, as a result of external events as well as from the actions (and inactions) of the project team and others elsewhere in the organisation. In order to optimise the chances of meeting the project's objectives, it is essential that the project team have a current view of the risks facing the project, including both threats and opportunities. For risk management, standing still is going backwards. This is reflected in Figure 3.1 by the two arrows leaving the Risk Review box. It is usually enough to repeat Risk Reviews throughout the life of the project, checking progress on planned actions, considering additional responses, identifying new risks and reassessing risks that are still current. However, sometimes it may be necessary to return to the start of the risk process and run through the earlier steps again. This may be required at key milestones or project gateways, or if there are significant changes in the scope or context of the project.

Post-Project Risk Review

As soon as the project has been completed, it is common for the project team to be disbanded and reassigned to their next project. This should not be done however without taking time to capture the lessons which can be learned from this project and applied to benefit similar future projects. It should be a routine part of the standard project management process to conduct a post-project review, but it is often omitted as discussed earlier in this chapter. Where it is conducted, then the organisation should ensure that it includes consideration of risk-related aspects. In the absence of a formal review at project level, the risk process needs to include a final Post-Project Risk Review step to ensure that useful information is not lost. Table 3.2 shows that most of the commonly used risk management standards and guidelines omit this vital step from the risk process, but we recommend its explicit inclusion to ensure that it is not overlooked.

Whether it is done as part of the wider project management process or specifically focused only on risk management, the Post-Project Risk Review step should consider a range of important risk-related questions, including:

- What were the main risks identified on this project (both threats and opportunities)? Do any of these represent generic risks that might affect similar projects in future?
- Which foreseeable threats actually occurred, and why? Which identified opportunities that could have been captured were missed, and why?
- Which issues or problems occurred that should have been foreseen as threats? Which unplanned benefits arose that should have been identified as opportunities?
- What preventative actions could have been taken to minimise or avoid threats? What proactive actions could have been taken to maximise or exploit opportunities?
- Which responses were effective in managing risks, and which were ineffective?
- How much effort was spent on the risk process, both to execute the process, and to implement responses?
- Can any specific benefits be attributed to the risk process, for example, reduced project duration or cost, increased business benefits or client satisfaction and so on?

Addressing these questions will ensure that the organisation gains the full benefit from undertaking the project, not simply producing a set of project deliverables, but contributing to organisational learning and knowledge.

Not 'One-Size-Fits-All'

It is clear that different projects are exposed to different levels of risk, so each step in the project risk management process must be scalable to meet the varying degrees of risk challenge. Scalable elements of the process include:

■ *Allocation of people to tasks.* In the simplest case the project manager may undertake all the elements of the risk process as part of their overall responsibility for managing the project, without using specialist risk resources. At the other extreme a complex risky project may require input from people with particular risk skills, and a dedicated risk team may be employed, either from within the organisation or from outside. Table 3.7 describes the roles and responsibilities for risk management for a larger project, and these may be combined and performed by fewer individuals on smaller projects.

■ *Methodology and processes used.* A low-risk project may be able to incorporate the risk process within the overall project management process, without the need for specific risk management activities. A riskier project may need to use a defined risk process, perhaps following a recognised risk methodology.

■ *Tools and techniques used.* The simplest risk process might involve a team brainstorm undertaken as part of a routine project team meeting, recording risks in a spreadsheet and monitoring actions through the regular project review meetings. The riskiest projects may require a wide range of techniques for risk identification, assessment and control, to ensure that all aspects of risk exposure are captured and dealt with appropriately.

■ *Supporting infrastructure.* The lowest-risk projects may require no dedicated risk infrastructure, whereas high-risk projects demand robust support from integrated toolkits with high levels of functionality. It is important to get the level of infrastructure right as too much 'support' can strangle the risk process and too little can leave it unable to function.

■ *Reporting requirement.* For some projects the reporting of risk exposure can be incorporated into the overall routine project reports, whereas others may demand a variety of specific risk reports targeted to the needs of different stakeholders. The aim is to ensure that each group of stakeholders gets timely risk information which is relevant to their interest in the project.

■ *Review and update frequency.* It may be sufficient on low-risk or short duration projects to update the risk assessment only once or twice during the life of the project. Other projects which are riskier or of longer duration may need a regular risk update cycle, say monthly or quarterly, depending on the project's complexity and rate of change.

Table 3.7   Roles and Responsibilities Within the Risk Process

Project Sponsor. Accountable for the overall project and for delivering its promised benefits, and as such can be considered to be the ultimate risk owner for the project. The Project Sponsor must ensure that resources and funds for risk management are provided to the project. The role of the Project Sponsor includes:

- Actively supporting and encouraging the implementation of risk management on the project.
- Setting and monitoring risk thresholds and ensuring that these are translated into acceptable levels of risk for the project.
- Attendance at risk workshops as required by the Project Manager.
- Identification of risks in their area of responsibility.
- Ownership of risks as required by the Project Manager.
- Reviewing risk outputs from the project with the Project Manager to ensure process consistency and effectiveness.
- Reviewing risks escalated by the Project Manager which are outside the scope or control of the project or which require input or action from outside the project.
- Taking decisions on project strategy in the light of current risk status, to maintain acceptable risk exposure.
- Ensuring that adequate resources are available to the project to respond appropriately to identified risks.
- Releasing 'management reserve' funds to the project, where justified, to deal with exceptional risks.
- Regularly reporting risk status to senior management.

Project Manager. Responsible for delivering the project on time, within budget and to the agreed level of quality such that the project's outputs will allow the promised benefits to be achieved. The Project Manager is accountable for the day-to-day management of the project. Part of this requires ensuring that the risk process is properly and effectively implemented. The role of the Project Manager includes:

- Determining the acceptable levels of risk for the project (in consultation with the Project Sponsor).
- Approving the Risk Management Plan prepared by the Risk Champion.
- Promoting the risk management process for the project.
- Participating in risk workshops and review meetings.
- Identification of risks.
- Ownership of risks as appropriate.
- Approving risk response plans and their associated risk actions prior to implementation.
- Applying project contingency funds to deal with identified risks that occur during the project.
- Overseeing risk management by subcontractors and suppliers.
- Regularly reporting risk status to the Project Sponsor and project board/ steering committee, with recommendations for appropriate strategic decisions and actions to maintain acceptable risk exposure.
- Highlighting to senior management any identified risks which are outside the scope or control of the project, or which require input or action from outside the project, or where release of 'management reserve' funds might be appropriate.
- Monitoring the efficiency and effectiveness of the process in conjunction with the Risk Champion.

(Continued)

Table 3.7 (Continued)   Roles and Responsibilities Within the Risk Process

Risk Champion. Responsible for overseeing and facilitating the risk management process on a day-to-day basis. Note that this might be a full-time role or a part-time role. The role of the Risk Champion includes:
- Preparing the Risk Management Plan.
- Facilitating the risk process, including risk workshops and risk review meetings.
- Creating and maintaining the Risk Register.
- Liaising with Risk Owners to determine appropriate risk responses.
- Ensuring the quality of all risk data.
- Analysing data and producing risk reports.
- Reviewing with Risk Owners the progress of risk responses and their associated actions.
- Advising the Project Manager on all matters relating to risk management.
- Coaching and mentoring team members and other stakeholders on aspects of risk management.

Risk Owner. Responsible for managing a specific identified risk. One Risk Owner is appointed for each risk by the Project Manager in liaison with the Risk Champion. The Risk Owner's role ceases once that risk is no longer active. A Risk Owner can be a member of the project team, a stakeholder who is not part of the project team, or a specialist from outside the project. The role of the Risk Owner includes:
- Developing responses to risks in the form of risk actions which they then assign to Action Owners.
- Monitoring the progress on their risk responses.
- Identifying secondary risks.
- Reporting progress on responses to the Risk Champion via the Risk Register.

Action Owner. Appointed by Risk Owners to perform one or more of the actions that make up a response to a risk. The role of the Risk Action Owner ceases once their action(s) has been completed. Several Action Owners may contribute to the response to one risk. The role of the Action Owner includes:
- Implementing agreed actions to support response strategies.
- Reporting progress on actions to the Risk Owner and recommending any other actions needed to manage the risk.
- Identifying secondary risks.

Project Team Members. Responsible to the Project Manager to ensure that the risk process is followed by themselves and others who report to them. The role of project team members includes:
- Participating actively in the risk process, proactively identifying and managing risks in their area of responsibility.
- Participating in risk workshops and risk review meetings as required.
- Providing inputs to the Project Manager for risk reports.

Other Stakeholders. All project stakeholders must be involved in risk management as appropriate. Stakeholders may be both causes of risks and the source of responses to risks. Some stakeholders might be classified as 'key stakeholders', and these will be required to participate actively in the risk process.

Once these scalable aspects are determined for a given project, they should be documented in the project's Risk Management Plan, as part of the Risk Process Initiation step (see above).

More than a Process

It is a common fallacy to think that risk management is just a process. Indeed, many of the risk standard and guidelines reinforce this impression, by focusing on the practical steps required to manage risk. This makes it easy to think of risk management as merely a combination of tools and techniques put together in a structured framework. This leads to the view that all the project team has to do is follow the process and risks will be managed effectively. Of course, a process is important, but it is not the whole story. Risk management processes are necessary but not sufficient. The truth is that there are several other significant factors in addition to the process that influence how well risk is managed on projects. These additional success factors are addressed in subsequent chapters. Perhaps the most important influence is the people aspects of managing risk, since risk is ultimately managed by people and not by robots or machines. The softer elements of risk psychology and risk attitudes are tackled in Chapter 5. Then there are a range of integration issues to be addressed to ensure that risk management is not conducted in isolation. Integration of risk management is required at two levels: within the project management process (addressed in Chapter 6), and more widely in the organisation (covered in Chapter 7).

But before dealing with these important elements that affect management of risk in all projects, we need to consider how to tackle risks in a particular class of projects that require special attention: complex projects. This is discussed in the following chapter.

Chapter 4

Managing Risk in Complex Projects

The Challenge of Complexity

Use of the VUCA abbreviation has become more widespread in business and general use in recent years, as the all-pervasive nature of Volatility, Uncertainty, Complexity and Ambiguity is recognised and accepted as normal.

- *Volatility* occurs when the nature, speed and size of change are both large and unpredictable.
- *Uncertainty* arises from lack of knowledge or an inability to determine the course of future events.
- *Complexity* is present when the outcome of an action cannot be predicted by simple analysis.
- *Ambiguity* means that key characteristics of a situation are not clear, or they can be interpreted in different ways.

Each of these four characteristics finds a particular relevance and application in the world of projects, where there has been a special focus on complexity for some time, driven by the increasing size and scope of projects. This is illustrated by the establishment in 2007 of the International Centre for Complex Project Management (ICCPM), with a mission to equip project professionals and organisations with the specialised skills, knowledge and ability to deal effectively with complexity in project environments (https://iccpm.com/).

DOI: 10.4324/9781003431954-4

Although its genesis was in the defence communities of Australia, UK, US and Canada, ICCPM has grown to become the main professional body in this area across all industry and government sectors, comprising a substantial network of global corporate, government, academic and professional organisations committed to the better management and enhanced success rate of complex projects.

Before considering further how to manage risk in complex projects, we need to be clear about the meaning of *complexity*. Many people confuse the terms *complex* and *complicated*; this is more than mere semantics as the two situations pose very different management challenges.

The Cynefin sense-making framework was developed by Snowden in 1999, distinguishing five distinct types of situations, as illustrated in Figure 4.1, namely Clear, Complicated, Complex, Chaotic, and Confused:

■ Clear situations are well-defined, with evident cause–effect links enabling deterministic prediction, rational analysis and effective management.

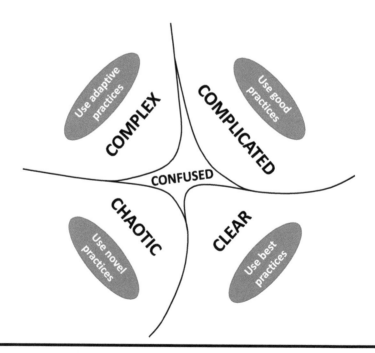

Figure 4.1 The Cynefin Framework (Adapted From https://thecynefin.co/about-us/about-cynefin-framework/)

- Complicated situations have many elements with more indirect cause–effect relationships, requiring more detailed analysis, often involving expert judgement, to define the way the elements interact.
- Complex situations are unpredictable, with unclear links between constituent parts which are often only understandable in retrospect when patterns may emerge.
- Chaotic situations have no apparent rules governing their behaviour, making it difficult or impossible to understand and manage what is going on.
- Confused situations appear to be unknowable and unpredictable since identifiable factors driving their behaviour are ambiguous or contradictory or absent.

When applied to the domain of projects, it seems obvious that a project should never be launched in situations which are Chaotic or Confused, and that projects that enter one of those states at some point during their lifetime need urgent attention if they are to proceed. Clear projects are usually simple and straightforward to manage to completion, with well-defined objectives, constraints that are understood and acceptable, and resource requirements that can be satisfied when required. Risk management in Clear projects can follow the standard risk process outlined in the previous chapter, to identify, assess and manage threats and opportunities proactively and effectively.

The main challenges in project management are posed with situations that are Complicated or Complex. However, the approach to understanding and managing risk in these two types of projects are quite different, arising from the distinct nature of the two situations. This is best understood by exploring the underlying roots of the two words.

- Complicated originates from the Latin verb *complicare*, meaning *to fold together*.
- Complex comes from the Latin verb *complectere*, which means *to knot/plait/entwine together*.

The difference can be illustrated by comparing an origami model with many folds (complicated) with a plate of cooked spaghetti (complex).

- A complicated origami model can be unfolded until it is only a flat sheet of paper, and then refolded by following the same process in reverse to recreate the same end product. The precise relationship between each part of the model is clear and unambiguous.

■ By contrast, if one end of a strand of spaghetti is pulled it is impossible to predict in advance what effect that might have on other strands, and if one or more strands are removed from the bowl it would be impossible to replace them to recreate the original conformation.

Similarly, a Complicated project may have many elements, but the relationships between those elements are defined such that the project can be decomposed into more detail for analysis and management, and then recombined to form the whole project. This ability to break a Complicated project into smaller parts enables many of the standard project management processes to be implemented without difficulty, including the management of risk. A reductionist approach works well here, perhaps involving the use of subprojects or work domains whose size is more amenable to analysis, followed by a synthesis to form a whole-project perspective.

A Complex project may also contain many elements, but the way they interact is not obvious. Behaviour of the project as a whole cannot be predicted by analysis of its component parts. Change in one part of the Complex project will have unforeseen (and perhaps unforeseeable) impacts elsewhere in the project, leading to emergent properties many of which are not susceptible to management using standard project processes or techniques. Risk in a Complex project is one such emergent property, as uncertainty arises in places where it had not been expected or predicted, requiring an adaptive approach that is quite different from the reductive methods used in Clear or Complicated projects.

This use of different approaches is also recommended by the Cynefin framework, as shown in Figure 4.1.

■ *Best practices* can be used in Clear situations where the requirement is understood and the most appropriate techniques are evident, applying well-defined and predetermined rules and approaches that are guaranteed to achieve the desired effect.

■ Where a situation is Complicated, analysis should indicate which approaches are most likely to work, so *good practices* can be applied, guided by expert judgement and experience.

■ Complex situations demonstrate emergent properties so they require the use of *adaptive practices* that can respond to a developing challenge.

The terms 'exaptive practices' or 'exaptation' are sometimes used, describing an approach where an existing practice is repurposed and used in a way that differs from its original design intent.

■ Situations that are Chaotic must be managed reactively, aiming to impose order and control where possible, and *novel practices* may be required that are developed specifically to do whatever is necessary.

■ In cases of Confusion, it is best to do nothing until the situation resolves into one of the other four types.

Having clarified the meaning of complexity, we can now turn to the management of risk in complex projects. This differs from the management of complex projects as a whole, with a focus on proactively finding and addressing those uncertainties that might arise in the course of the project. It is true that complex project management itself will address riskiness by seeking to understand the causes of complexity in a particular project and designing the project approach accordingly. Management of risk within a complex project is more targeted and specific in intention, focus and application.

Identifying Risk in Complex Projects: Futures Thinking

We first need to consider what types of risk might occur in complex projects. These projects are different from non-complex projects, since they are inherently ambiguous and unpredictable. However, complex projects are subject to the same types of risk as all other projects, in addition to emergent risks that arise during project execution. This means that complex projects need to adopt two parallel approaches to managing risk:

1. The traditional risk management process outlined in Chapter 3 can be used for typical risks that are common to all projects;
2. Emergent risks need something more flexible and generic which can respond quickly and appropriately when the nature of the risk becomes apparent.

Early attempts to manage emergent risk in complex projects took a systems-based approach, treating projects as complex adaptive systems, and using insights and techniques from systems analysis to model the combined effect of complexity and uncertainty in order to identify risk (Williams, 2002; Barber, 2003). At their simplest, these might involve building an influence diagram or entity relationship diagram to represent the whole project as a system of interrelated elements, then flexing the model by changing its variables to expose areas of instability and uncertainty. The approach can be developed further into full systems dynamics modelling, with use of powerful analytical techniques that can reveal counter-intuitive behaviour in areas of the project. While these approaches offer unique insights into where risk might arise in a complex project, they require a degree of expertise and experience that is not always available (or affordable), supported by sophisticated tools.

Recent guidance from the UK Association for Project Management (Association for Project Management, 2024) includes a number of project characteristics that are indicative of complexity, and which might usefully be explored when seeking to identify risk. These include:

- *Project size,* including overall duration and budget, solution elements, number of stakeholders, project team size, supply-chain components, etc.
- *Element uniqueness,* describing the extent to which first-of-a-kind project components are being developed specifically for this project.
- *Environment uniqueness,* considering economic, geographical, regulatory, socio-political or cultural aspects within which the project is being developed.
- *Novelty,* which covers the degree to which experience and expertise possessed by the project team and performing organisation is relevant to the challenge posed by this particular project.
- *Team capacity and capability,* addressing organisational and structural aspects of the project that could affect decision-making or performance.
- *Stakeholder expectations,* which are more likely to overlap and conflict for a complex project with ambiguous and emergent properties.

Other factors that drive complexity in projects include requirements clarity/stability/volatility, solution urgency, strategic importance, and degree of change (Hass, 2009).

The question remains how to explore these dimensions of complexity in a way that indicates the level of risk faced by the project as a whole, and that points the way towards managing that risk effectively. Analytical techniques such as quantitative stochastic modelling or business informatics analysis may be useful in this context, but their application is limited and practical guidance is lacking.

Fortunately, other techniques which are more generally accessible can be applied to the challenge of identifying risk in complex projects. These can be described under the heading of *Futures Thinking* (Government Office for Science, 2021), which involves four basic steps:

1. Define Preferred Future(s);
2. Identify Possible Futures;
3. Analyse to find Probable Futures;
4. Manage future proactively to make Preferred Future(s) more/most probable.

This application of Futures Thinking to identifying and managing risk in complex projects is an example of 'exaptation', as described in the previous section, since the technique was not originally intended or designed for this purpose. It is typically used to generate scenarios for long-range strategic planning.

One or more Preferred Futures for the project are described using standard approaches to objectives-setting or goal elaboration. Various techniques can then be used to identify Possible Futures, including the following:

■ *Scenario analysis.* This generic term is used for a variety of techniques that can be used to generate possible future states. These include: *quantitative models* that apply ranges to variables (typically worst-case/likely-case/best-case), with or without use of probabilistic analysis; *gaming exercises* (sometimes called 'war-gaming') where teams compete to select values for variables that maximise performance outcomes, leading to identification of a number of possible 'winning combinations'; *event-driven scenarios*, in which the effect of a range of disruptors is considered using structured multiple-layer 'What if?' questioning; *alternative futures scenarios*, where permutations of external macro-level forces and boundary conditions are explored in a brainstorming setting, using open questions to stimulate creative thinking.

■ *Trend analysis.* This uses mathematical techniques to analyse historical data and predict potential trends. It starts with systematic data-gathering for relevant similar situations over a period of time, either within this complex project or in others that have similar character-istics and parameters. Best-fit techniques are used to generate possible trend lines to extend past performance into the future, either linearly or using more complex functions, recognising that predictions become more uncertain the farther out they extend. Trend analysis is likely to be an area where Artificial Intelligence could be valuable, finding hidden trends and patterns that are not immediately obvious to standard predictive techniques.

■ *Horizon scanning.* This is essentially a structured form of Delphi group analysis that focuses on future uncertainty, using subject-matter experts to identify potential developments within a predefined set of time horizons (Institute of Risk Management, 2018). These horizons are often simply stated as short-term (events and circumstances with immediate impact and where action can be taken now), medium-term (trends that are currently evident or likely to develop which should inform strategic positioning), and long-term (more speculative trends that are not currently visible but which are expected to grow in significance).

■ *Field Anomaly Relaxation (FAR).* This somewhat exotic technique was initially developed in 1974 by Rhyne as a structured method for generating feasible futures (Rhyne, 1974). It involves describing an initial view of possible futures and their main characteristics ('sectors'), identifying alternative states for each sector ('factors'), and constructing a matrix of all possible configurations. Factor pairs that are inconsistent or mutually exclusive are eliminated ('relaxed'), and the remaining subset of futures are arranged into a smaller group of related future states, together with transitions between them, to create a 'futures tree'. Despite its methodical nature, FAR has not been widely used in practice, but it remains a powerful option to be considered for some particularly complex situations (Coyle, 1997).

Having generated a range of Possible Futures for our complex project, their likelihood of occurrence is usually estimated using expert judgement, captured via brainstorms, Delphi groups, appreciative enquiry, or structured questioning techniques, allowing the ranking of Possible Futures, including the position of the Preferred Future(s). Where at least one of our Preferred Futures is not among the most likely, we can make risk-based decisions and take risk management actions to influence the future proactively, where possible, by changing some Possible Futures and introducing new ones. This process is illustrated in Figure 4.2 (before) and Figure 4.3 (after).

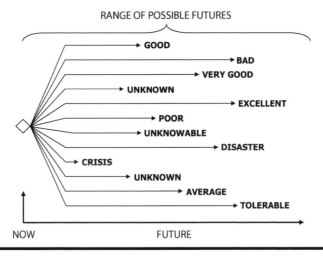

Figure 4.2   Range of Possible Futures (Before Action)

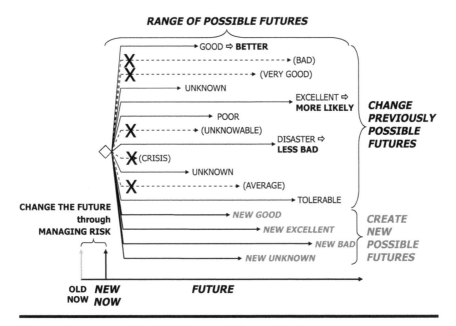

Figure 4.3   Range of Possible Futures (After Action)

Figure 4.2 shows that as the project proceeds, there are a range of Possible Futures that lie ahead. These might result in outcomes that are good, bad or indifferent to different degrees, and of course some outcomes are either unknown or unknowable. Depending on how the project progresses, each of these outcomes might occur at a different point in the future.

Figure 4.3 illustrates how taking risk-based decisions and risk management actions can change the future (Murray-Webster & Hillson, 2021). As we move to a new 'Time Now' following implementation of such decisions and actions, some of the previously Possible Futures become infeasible and are excluded from the set of futures that might now occur (shown by an 'X' in the figure). Unfortunately, while we aim to influence outcomes that are less desirable, we are not always to predict the consequences of our actions precisely, so we may end up excluding Possible Futures that would have been good. We are also likely to have unknown effects that are not visible to us.

The nature and degree of other Possible Futures is modified, sometimes in a desirable way (good outcomes become better, excellent outcomes become more likely, disastrous outcomes become less bad, positive outcomes might occur sooner), while others remain unchanged as they are not amenable to influence. In addition to changing the likelihood and impact of the set of Possible Futures that existed when we planned and implemented our risk management decisions and actions, the lower part of Figure 4.3 shows that our actions will create a new set of Possible Futures including some that are good, bad, indifferent or unknown.

In this way, risk management allows proactive management of the future in a way that offers a powerful response to the challenge of emergent risk and uncertainty in complex projects. But what kind of risk-based decisions and risk management actions should we consider as we seek to influence the range of futures that might lie ahead?

Managing Risk in Complex Projects: Adaptive Resilience

Complex projects contain some risks that are typical of all projects, and these can be managed using the standard risk response strategies as outlined in the previous chapter. These include Escalate, Avoid/Exploit, Transfer/Share, Reduce/Enhance, and Accept. These all allow proactive planning of appropriate treatments and actions, many of which can be implemented in advance of the risks' possible occurrence.

These routine risk responses are however not relevant for managing emergent risks in complex projects, which appear too late for such proactive planning and action. A different approach is required here, which provides the necessary protection for objectives while maximising the chances of success. In the broader business context, this is the purpose of *business continuity management* (BCM), which aims to preserve and maintain the capability of the organisation to deliver products and services at the required level of performance following a disruptive event or set of circumstances. BCM allows use of advanced planning to prepare the organisation to maintain or rapidly recover business functions in the face of challenges that lie outside the normal business parameters.

A similar approach can be adopted for management of emergent risk in complex projects, which we can term *project continuity management*. (This repurposed application of BCM to the complex project environment is an example of an 'exaptive practice', as described earlier.) Here the goal is to build resilience within the project as a whole in such a way that it retains flexibility to respond appropriately to any challenge that may emerge in future. We call this *adaptive resilience.*

Much has been written about resilience in recent years, crossing a wide range of domains from organisational performance to personal development and mental health, and much in between. In the business and organisational context, the focus on resilience initially arose from the recognition that for an organisation to be merely robust was insufficient to meet situations where the stress imposed by external forces exceeded the ability of the organisation to maintain performance at acceptable levels (Seville, 2016). Where robustness proves inadequate and organisational performance begins to suffer, resilience steps in to prevent complete collapse and ultimate failure, allowing the organisation to recover when the immediate crisis is over. The key characteristic of a resilient organisation is the ability to return to acceptable levels after experiencing performance degradation.

While organisational resilience is quickly maturing into an established discipline, its application to complex projects needs some additional features. This is due to the emergent nature of complex projects which can rapidly produce critical situations with limited time to respond. It is in these circumstances that we need *adaptive resilience.* A base level of resilience can provide a foundation to protect the project, but this may be insufficient when significant risk and uncertainty emerge. Here, a degree of flexibility is required in the way pre-planned responses are focused and implemented. Measures that might have normally been expected to provide protection from disruption may need adjustment, and new measures may need to be developed quickly. This could require a set of contingency plans that offer additional levels of resilience beyond what was initially planned for.

Typical frameworks for building resilience aim to return the organisation to its prior state, while adaptive resilience goes further, seeking not only to 'bounce back' but to 'build back better'. However, this slogan begs a question: What is meant by 'better'? If the goal is to build back the same as previously but just operating more efficiently (perhaps characterised as 'faster, smarter, cheaper'), then one might question whether this is the best possible outcome. An alternative suggestion might be that 'better' should not merely mean to build back to the same objectives with more efficiency, but might mean working towards different objectives which in themselves are better than the previous ones. Where standard approaches to resilience might aim to 'bounce back', the goal of adaptive resilience is instead to be sufficiently flexible to 'bounce forward'. This might include a strategic review of vision, mission and objectives, performed in the light of the new situation that exists following the removal of the disruption.

Practical implementation of adaptive resilience in complex projects will depend on the specific nature of each project, but some common features are needed, within an overall structured approach to project continuity management.

Practical Project Continuity Management

The discipline of *business continuity management* (BCM) is well established and understood, and it is practised widely in organisations across a range of industries and in different countries. It includes a mature process, with development of a business continuity plan (BCP) that outlines measures of performance that indicate that significant deviation is either occurring or imminent. These include both Key Performance Indicators (KPIs) and early-warning indicators. The BCP also describes proactive and reactive responses to disruptive circumstances, with enough detail to show who does what when, where and how. Necessary resources to implement these responses are identified and planned, including pre-positioned assets, and the BCP is exercised, reviewed and updated on a regular basis to ensure that it remains effective for all foreseeable situations, as well as providing flexible protection to cope with the unforeseen.

Each of these elements of BCM can be implemented at project level, with adjustments to the level of focus and action. This would probably be unnecessary for the majority of routine projects, where standard project processes such as risk management and change management should prove adequate. However, a more developed approach is needed for complex projects, where emergent risk can present existential challenges for both the project, the organisation and some members of the wider stakeholder network. In these cases, a structured approach to *project continuity management* is needed. This should include development of a Project Continuity Plan with KPIs and early-warning indicators, with planned proactive and reactive responses, and provision of necessary resources. This plan should be tested and updated regularly throughout the project lifecycle.

To be implemented effectively, project continuity management for complex projects requires a number of enabling factors to be in place, including people, project, technical/contractual, and business/organisational. Building adaptive resilience in each of these areas is discussed in the following sections.

- **People**. A resilient project requires resilient individuals to form the project team. Individual resilience is partly a result of the innate characteristics of the people who make up the project team, including emotional literacy and positive attitudes (Hillson & Murray-Webster, 2007). There are also a number of qualities of the team together which contribute to the creation of adaptive resilience, including an empowered team with effective leadership and a supportive culture (Murray-Webster & Hillson, 2008).

  To ensure the degree of flexibility demanded by adaptive resilience, both individuals and the project team as a whole also need a high level of *cognitive readiness* (Belack et al., 2019). This describes the mental skills, knowledge, abilities and motivations that are necessary to establish and sustain outstanding performance in a complex, unpredictable and fast-changing environment. It encompasses aspects of cognitive intelligence, emotional intelligence and social intelligence. Among the personal characteristics of resilient individuals and teams, cognitive readiness is particularly important for the effective management of emergent risk and uncertainty in complex projects. High cognitive readiness enables individuals and teams to respond rapidly, appropriately and effectively to unexpected situations that arise without warning. This requires them to recognise patterns in unfamiliar situations, identify candidate solutions that may only have partial efficacy and modify these to address what has emerged, then to design and implement plans of action based on these modified solutions.

■ **Project**. Resilience in a project arises from the degree to which the project has firm and well-defined foundations, scope and boundaries. This is particularly important in the case of complex projects, where it may not be possible to predict the ultimate outcome or the path that may be followed on the way to project completion. Clear definition of project objectives is essential, as these will form the touchstone against which project progress and direction is assessed: are we moving towards or away from our goal, and are we maintaining a necessary level of performance to reach it?

The project will require flexible project management processes which can be adapted as necessary to meet the changing situation that will inevitably arise from the complex nature of the project. Among these processes, strong change management is particularly important, as it will enable the project to respond to generic volatility and specific unexpected factors, both in the external and internal environments, while maintaining a clear focus on the ultimate goal. Perhaps unsurprisingly, risk management capability is also a vital driver of project resilience, but the use of risk techniques must be flexed to respond to emergent uncertainty, rather than being bound to prescriptive risk processes and methodology. This must include appropriate levels of both risk budget and contingency, with clear rules for their use. Attention to rapidly and regularly identifying and learning lessons for improvement will ensure that the project can respond appropriately to emerging conditions. The use of evolutionary and adaptive development approaches also introduces helpful flexibility.

■ **Technical/Contractual**. Each complex project is different and unique in its technical solution and challenges, but technical resilience can be strengthened by the use of iterative or adaptive development and design redundancy where possible. The details of this depend on the specific technical nature of the project itself.

Similarly, contractual resilience requires clear accountabilities to be defined within the contract, as well as terms & conditions that can be adapted as necessary to meet the emergent needs of the project as it progresses. Flexibility is enhanced by avoiding so-called boiler-plate sections within contracts, with careful use of constraining words such as 'must', 'will', 'cannot', 'always' or 'never'.

■ **Business/organisational**. The organisational context within which complex projects exist is a significant factor in the extent to which resilience is possible. In particular, where a business has a shared corporate culture and grounded values that are widely communicated, understood and accepted, it will be laying the foundation for its projects to demonstrate adaptive resilience in the face of disruption. A strong approach to stakeholder engagement is also helpful in ensuring that complex projects have the type of supportive environment that allows them to respond flexibly to change while maintaining focus on their ultimate goal.

Current Guidance on Risk and Complexity

This chapter has focused not on managing complex projects, but on managing risk in complex projects. The approaches described here are necessarily generic because there is no current consensus on how risk in complex projects could be or should be identified, assessed and managed in detail. One might expect professional associations to provide guidance on how to manage risk in complexity, but this is still at an early stage, and to date their contributions have remained largely exploratory. Among the professional bodies, two stand out as candidates to provide guidance for the management of risk in complex projects:

- *International Centre for Complex Project Management (ICCPM)*. This has provided much useful material on management of complex projects as a whole, but does not currently offer much practical guidance on managing risk. For example, the results of an ICCPM international roundtable consultation in 2020 were published in a report with the subtitle 'Rethinking risk, opportunity and resilience' (International Centre for Complex Project Management, 2021), but much of this describes the problem space in theoretical and philosophical terms, with an academic examination of alternative approaches to understanding emergence, and limited discussion of practical steps that can be taken to respond appropriately to emergent risk. ICCPM has a Special Interest Group (SIG) dedicated to managing risk in complexity (ICCPM MRC SIG), and this has produced some broad guidance on approaches, methods and tools that can aid understanding of complex risks, especially where these are interconnected, as well as outlining how to improve take up of approaches to management of risk in complex projects (https://iccpm

.com/managing-risk-in-complexity-sig-progress-update-sept/). Access to more detailed outputs from this SIG is limited to ICCPM members in a dedicated ICCPM MRC SIG Forum.

■ *Institute of Risk Management (IRM).* While the scope of IRM is necessarily broad, it has a number of Special Interest Groups (SIGs) focusing on specific aspects of the risk challenge. These include a Risk & Complexity SIG (https://www.theirm.org/join-our-community/special-interest-groups/risk-and-complexity/), but at the time of writing this has not yet addressed the challenge of managing risk in complex projects. The material produced to date focuses on understanding the nature of complexity in general, and its relationship with risk and uncertainty in particular, but the application area of interest of this group is much wider than projects so targeted guidance for complex projects may not be a top priority.

In the absence of consensus and guidance from the relevant professional bodies on good practice for managing risk in complex projects, this chapter recommends the use of *Futures Thinking* to identify areas of risk, and the development of *adaptive resilience* through structured *project continuity management* to respond to emergent risk as the project proceeds. When more detailed guidance becomes available, it is likely that these broader approaches will provide an appropriate context and environment for implementation of specific risk management methods, tools and techniques to handle emergent risk in complex projects.

Chapter 5

Risk and People

Much of what is written about risk management concentrates on processes, and indeed that was the focus of Chapter 3. This might lead us to conclude that process is all that really matters when seeking to manage risk. If an individual, team or organisation pays attention to ensuring execution of a good robust risk process, supported by the Three Ts of Tools, Techniques and Training, then surely that is all that can be expected of them. They believe that faithfully following the process will inevitably lead to success, and the main measure of risk management success is equated by some with mere compliance to the approved risk process.

While it is certainly true that the Three Ts are important, they are not the whole story. An effective risk process is necessary but not sufficient. It is important to remember the purpose of the risk process. In fact, the reason for undertaking the risk management process is not (or should not be) simply to comply with the process. It is axiomatic to say that the risk management process exists to allow risk to be managed. And management of risk is only achieved by people actually using the results of the risk process to inform and modify their decisions, behaviour and actions. Unfortunately, there are many factors that affect the extent to which people are prepared to use risk results in practice, including the way they perceive the degree of risk that they face and their ability to influence it. These people-related factors need to be understood and managed if risk management is to fulfil its promise and deliver improved performance.

Consequently, it is essential for anyone who is committed to managing risk effectively in their projects or business to be aware of the *people aspects*

DOI: 10.4324/9781003431954-5

of risk management, and to actively manage these aspects alongside the risk process. This chapter explores the human side of risk management, focusing on how people respond to uncertainty and risk, and how this response affects their judgement and behaviour.

Understanding Risk Attitude

A number of terms are used to describe how people respond to uncertainty and risk, including risk attitude, risk appetite, risk threshold, risk tolerance, risk capacity, and so on. We'll return to what these mean and how they relate together shortly in this chapter, but first we address the key people factor when it comes to risk – risk attitude.

We have already seen in Chapter 1 that risk can be simply described as 'uncertainty that matters', and this proto-definition leads to a number of useful insights and approaches. It is however important to recognise that this phrase tells us something significant about the softer side of risk management. Both 'uncertainty' and 'mattering' are subjective terms, driven by the perceptions of the individual or group that is considering the risk. The essential task of risk assessment is to answer two key questions: 'How uncertain is it?' and 'How much does it matter?' The answers will drive which response strategy is selected and what actions are taken (if any). In many (most?) cases there are no unambiguous 'right answers' to these two questions, and different people will reach different conclusions, some regarding a particular risk as very uncertain and mattering a great deal, while others will consider the same risk as less uncertain or insignificant.

The term 'attitude' can be defined as 'a chosen response to a given situation', and this is also driven by perception, since there are two key questions to be answered: 'What are the characteristics of the situation I/we face?' and 'How should I/we respond?' How these questions are answered will determine the attitude adopted by an individual or group towards a particular set of circumstances, and a range of responses are possible, without there necessarily being a single 'right answer'.

Combining the two terms 'risk' and 'attitude', and noting the importance of perception in both cases, we can generate a working definition of 'risk attitude' as:

> A chosen response to uncertainty that matters, influenced by perception

This definition contains several key aspects of risk attitude which are important in understanding how it should be managed:

- *Chosen.* Risk attitude is not predetermined or fixed, but it is adopted. Repeated choice can result in habituation which can appear to be constant, but experience demonstrates that even a firmly entrenched attitude can be overturned where necessary.
- *Response.* Risk attitude does not exist in a vacuum, but is a response to something specific, in this case to risk. It is often not possible to say precisely what risk attitude will be adopted until a particular risk situation is encountered.
- *Perception.* There are a number of influences that can affect which risk attitude is actually adopted, and these operate through their effect on perception of the risk.

Another important feature of risk attitude is that it does not just apply to individuals. Risk attitudes are also exhibited by groups of various types, including project teams, management boards, review bodies, user groups and so on. In fact, risk attitudes also exist outside the workplace, and can be seen in families, local communities, clubs, sporting teams, charities and so on, as well as more widely in society and at national and international levels. Our focus here is on risk management within the project space, so we will concentrate our attention on risk attitudes among project stakeholders.

One further characteristic of risk attitude needs to be explored before we can move on to consider how these attitudes can be managed in the context of project risk management. For any given uncertain situation, a range of different risk attitudes can be adopted, ranging from very cautious to very welcoming. This is reflected in Figure 5.1, which depicts the *risk attitude spectrum*. The spectrum indicates that, faced with a particular risky situation, people can respond in a variety of ways. Each of these four groups of responses does not represent a unitary risk attitude, but it is simply a subsection of the overall spectrum.

- Some individuals and groups will be uncomfortable to a greater or lesser extent in the presence of uncertainty, and may feel anxious, intimidated, afraid, cautious or restricted. This is labelled as *risk-averse*.
- Others will have a very different reaction, enjoying the uncertainty, seeing it as a challenge against which they can pitch their wits and

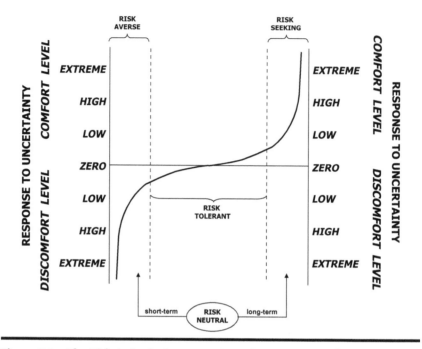

Figure 5.1   The Risk Attitude Spectrum (Based on Murray-Webster and Hillson, 2008)

demonstrate their abilities, providing a stimulus to creativity and innovation. This position is termed *risk-seeking*.

■ Still others may have no strong reaction, viewing the risk as an entirely normal and acceptable part of life, something which they can take in their stride without the need for any particular response. We call this *risk-tolerant*.

■ A fourth group may take a time-based view of the risk, being concerned to minimise their exposure in the short term while being prepared to take more of a chance in the longer term if there might be an advantage in doing so. This more nuanced response is called *risk-neutral*.

These four risk attitudes are described in more detail in Table 5.1.

It is perhaps natural for people to want to place themselves on the risk attitude spectrum, and label their own risk attitude as one of the four main options, but this is actually a complex question. When considering where an individual or group resides on the spectrum, most people will not be able to locate a single position that represents their risk attitude all of the time. Our experience in different risky situations is more variable, and we find ourselves being

Table 5.1 Risk Attitude Definitions and Characteristics

Term	Definition	Characteristics
Risk-averse	A conservative risk attitude with a preference for secure payoffs	Risk-averse individuals and groups are practical, accepting and valuing of common sense. They enjoy facts more than theories, and support established methods of working. They may feel uncomfortable with uncertainty, with a low tolerance for ambiguity, and be tempted to seek security and resolution in the face of risk. They may also tend to over-react to threats and under-react to opportunities.
Risk-seeking	A liberal risk attitude with a preference for speculative payoffs	People who are risk-seeking are adaptable and resourceful, enjoy life, and are not afraid to take action. They may underestimate threats, seeing them simply as a challenge to be overcome. They might also overestimate the importance of possible opportunities, wishing to pursue them aggressively.
Risk-tolerant	A balanced risk attitude with no strong reaction to uncertain situations	Risk-tolerant individuals and groups are reasonably comfortable with most uncertainty, accepting it as normal, and taking it in their stride with no apparent or significant influence on your behaviour. They may fail to appreciate the importance of threats and opportunities, tending to be reactive rather than proactive. This may lead to more problems from impacted threats, and loss of potential benefits as a result of missed opportunities.
Risk-neutral	An impartial risk attitude with a preference for future payoffs	People who are risk-neutral are neither risk-averse nor risk-seeking, but rather seek strategies and tactics that have high future payoffs. They think abstractly and creatively and envisage the possibilities. They enjoy ideas and are not afraid of change or the unknown. For both threats and opportunities, they focus on the longer term and only take action when it is likely to lead to significant benefit.

cautious in some circumstances, adventurous at other times, and sometimes tolerant of uncertainty or taking a time-based view. Our risk attitude at any given time is driven essentially by two factors: the external environment or situation, and the internal environment within us as individuals or as a group.

Similarly, if someone is asked where they should be on the risk attitude spectrum, and whether there is a 'right attitude' in a given situation, there is no single answer. Different risk attitudes are appropriate in different settings, depending on the objectives which are being pursued. So, for example, there is no one 'best risk attitude' for a project manager, since they may need to be risk-averse at times, for example, when a customer seeks to impose a major scope change, whereas it may be necessary for them to act in a risk-seeking manner at other times, perhaps when requesting additional project resources from senior management.

Influences on the Risk Attitude Spectrum

Although positioning on the risk attitude spectrum is not fixed, there are a range of influences that affect where an individual or group is placed, at least initially in the absence of active management. These influences act through affecting our perception of risk, with three major types of influence, known as the 'triple strand', illustrated in Figure 5.2. This is made up of *conscious factors, subconscious factors* and *affective factors*. While the three parts of the triple strand overlap and interact in complex ways, it is helpful to tease out each of the three elements so that they can be examined and understood.

■ *Strand 1 – Conscious factors.* These are the visible and measurable characteristics of a particular risky situation, based on our rational assessment. We also take account of situational factors such as whether we

Figure 5.2 The Triple Strand of Influences on Risk Attitude (Adapted From Murray-Webster and Hillson, 2008)

have done anything similar before (familiarity), the degree to which we have control of the situation (manageability), or how soon the situation is expected to affect us (proximity).

■ *Strand 2 – Subconscious factors.* These include heuristics and other sources of cognitive bias. Heuristics are mental shortcuts based on our previous experience. Some heuristics help us to reach an appropriate position quickly, while others can be misleading. Unfortunately, because heuristics are subconscious, their influence is often hidden, and they can be a significant source of bias. Common heuristics include memory of significant events (availability), or the conviction that we already know the right answer (confirmation trap).

■ *Strand 3 – Affective factors.* These are gut-level visceral feelings and emotions which tend to rise up automatically or instinctively in a situation and influence how we react. Fear, excitement or attraction can lead us to adopt risk attitudes which a more rational assessment might not consider.

The influences in the triple strand interact together to affect perception in two important ways: how people perceive a particular risky situation, and what they perceive as the right way to respond to it. By appreciating how the triple strand factors drive our perception of risky situations, we will understand better why we adopt different risk attitudes. This will help us to manage our attitudes to risk proactively so that we make good decisions, select appropriate responses and improve our management of risk.

Risk Attitudes and the Risk Process

While it is undoubtedly true that decisions in general should be made in a risk-aware manner, the human aspects of risk management also have a specific relevance to how risk is managed in projects. Indeed, risk attitudes exert a powerful influence over almost every element of the project risk management process. This is summarised in Table 5.2, which takes several key points in the risk process and shows how people with different risk attitudes might behave.

The notable differences are between the two extremes of the risk attitude spectrum, namely individuals and groups who are strongly risk-averse or those who are strongly risk-seeking. However, we must remember that risk attitude is not fixed, but it is influenced by the triple

Table 5.2   Influence of Risk Attitude on Key Points in Risk Process

Process step	Risk attitude			
	Risk-averse	*Risk-tolerant*	*Risk-neutral*	*Risk-seeking*
Risk process initiation	Low risk threshold, seeking to minimise level of risk to which the project or organisation is exposed	Medium-to-high risk threshold, prepared to accept a level of risk exposure as 'normal business'	Medium-to-high risk threshold, prepared to take risk now in order to achieve payback or advantage later	High risk threshold, prepared to take more risk in order to gain associated benefits
Risk identification	Tendency to identify many threats, but to ignore opportunities, driven by concern that opportunities may distract attention from management of threats	May treat risk identification as unimportant, since risks are accepted as a routine part of working on projects, leading to failure to identify risks	Focus on identifying risks with longer-term impacts, possibly even missing short-term project risks in favour of those affecting later phases or post-project operations	Tendency to play down threats and focus on opportunities, driven by desire to take more positive risk in order to maximise challenge and potential benefits
Qualitative risk assessment	Overestimation of threats in terms of both probability and impacts, and underestimation of opportunities	Many (most?) risks assessed as low probability and low impact	Assessments driven by proximity (time horizon), with higher proximity risks assessed as being more likely and/or bigger impact	Overestimation of opportunities in terms of both probability and impacts, and underestimation of threats
Risk response planning	Selection of aggressive and proactive strategies for threats, and tendency to accept or ignore opportunities	Preference for accepting risks	Response strategies driven by proximity, being more aggressive towards near-term risks and accepting risks where the potential impact is further into the future	Selection of aggressive and proactive strategies for opportunities, and tendency to accept or ignore threats
Risk response implementation	Conscientious implementation of agreed actions for threats, driven by desire to avoid or reduce risk exposure as much as possible, coupled with inattention to actions directed towards opportunities	Tendency to treat risk actions as low priority, to be implemented only if/when 'genuine project tasks' are completed	Focus of actions on high-proximity risks where impact could occur in near term	Tendency to ignore or postpone agreed actions targeting threats, and concentrate on actions aimed at exploiting or maximising opportunities

strand factors, so any particular individual or group might exhibit different risk attitudes and hence different behaviours at different times during the risk process or the project lifecycle.

Risk and Decision-Making

While the discussion on definition and characteristics of risk attitude is interesting in itself, it is important to know why this is important in the context of managing risk on projects. Every step in the risk process requires decisions to be made, and each of these decisions is influenced by our attitude to risk, shaped in turn by our perception. Examples include the following:

- Which objectives should be included within the scope of the risk process? What is at risk?
- What are our thresholds for acceptable risk exposure for the project as a whole? How much risk is too much risk?
- What criteria will we use for prioritising risks? How high is 'High'?
- How shall we respond to identified risks? Is it appropriate to do nothing and take the risk, or should we take action, and if so, what?

In general terms, our perception of risk is important because it affects our ability to make decisions. All human endeavour involves making decisions at all levels, including personal, private, professional, public and political. Decision-making has two key characteristics: it is *risky* and it is *important*.

- *Risky*. We have to make decisions where the situation is uncertain, often unknown and sometimes unknowable. If there is no uncertainty about what should be done, then no decision is needed ('Just do it!').
- *Important*. A decision is only required where there is more than one possible outcome, and the 'right decision' may not be evident. But a decision is also only required in situations where the result matters. If there is no significant consequence arising from alternative decision outcomes, then why bother deciding?

These two characteristics mean that decisions and decision-making are about *uncertainty that matters* – features they share with risk itself.

Consequently, decisions should be made in the light of risk, and assessment of the risk exposure associated with the various possible decision outcomes should form an intrinsic part of the decision-making process. This is where risk attitudes become important to decision-making, since they determine how we choose to respond to the perceived level of risk, and so exert a significant effect on both the decision process and the final decision outcome. Risk attitudes operate in the decision-making context at both individual and group levels, adding complexity to an already-difficult situation.

However, risk attitudes are not the only risk-related influence or input to decision-making. A number of other factors come into play, as illustrated in Figure 5.3 (see also Murray-Webster & Hillson, 2021). The decision-making process starts when one or more *decision-makers* are required to consider a set of *decision information* and select a *decision option* from among several alternatives. Decision information includes the objectives of the decision, the context within which the decision will be made (including assumptions, constraints, dependencies, stakeholder details, requirements for timing or cost or other desired outcome parameters), and the available decision options.

Given the nature of the decision information, decision-makers will have an inherent *risk appetite* which reflects their tendency to take risk in order to achieve their decision objectives. That internal tendency is expressed in quantified and measurable risk thresholds that represent upper and lower limits of

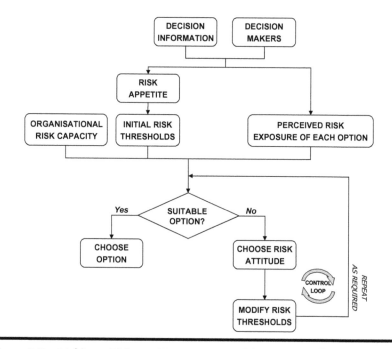

Figure 5.3　Making a Risky and Important Decision (From Hillson, 2022)

acceptable uncertainty against each decision objective. *Initial risk thresholds* are often set intuitively, based on a degree of gut feel about how much risk is acceptable in the context of this particular decision.

Two other key risk factors must be taken into account by decision-makers as they consider the alternative decision options: the *organisational risk capacity* (the overall ability of the organisation to bear risk) and the *risk exposure* of each option (which is coloured by the perception of the decision-makers).

At this point in the decision-making process, decision-makers have three key inputs:

- Measurable risk thresholds that express their inherent risk appetite (answering the question 'How much risk do we want to take in this decision?')
- Organisational risk capacity ('How much risk are we able to bear overall?')
- Perceived risk exposure of each decision option ('How much risk is associated with each option?')

This might enable them to make a decision on how to proceed, if the risk exposure of one decision option clearly lies within the risk thresholds and below the overall organisational risk capacity. However, if there is no unambiguously clear single suitable decision option, all is not lost. This is where risk attitude comes into play. We've seen that risk attitude is not fixed but it is variable, and it can therefore be chosen and changed intentionally if a different risk attitude might produce a more desirable outcome.

In the situation where no decision option is suitable, decision-makers have a choice over whether to modify the risk thresholds that they apply to the decision. The initial risk thresholds used in the decision-making process are based on the unmodified inherent risk appetite of the decision-makers, defining how much risk they are prepared to take in relation to this decision. But if these risk thresholds result in none of the available decision options being acceptable, decision-makers may consider it appropriate to adopt a different risk attitude which would change where risk thresholds are set. For example, if the inherent risk appetite for a given decision was low and initial risk thresholds were tight, decision-makers may wish to intentionally choose to take more risk by adopting a more risk-seeking attitude, allowing them to redefine risk thresholds more loosely. If one or more of the available decision options falls within these *modified risk thresholds*, they can now be considered where previously they would have been rejected.

In this way, risk attitude provides a control point in the decision process by offering the ability to intentionally choose to take more or less risk than would be acceptable if the only influence was unmodified risk appetite, allowing previously unacceptable options to be considered.

Managing Risk Attitudes

The variable nature of risk attitude, which can exist at any point across a wide spectrum, which differs from one individual or group to another, and which is chosen but influenced by a broad range of factors, presents a significant management challenge. If we want our management of risk to be effective, we need to make appropriate risk-based decisions within our project. We've seen that there are a number of risk factors at play when making decisions that are risky and important, and risk attitude acts as a vital control point in the process since it is variable and can be chosen and changed. But how can we proactively *manage* our own risk attitude and that of others in or project in order to optimise project decision-making?

The Seven A's model (Figure 5.4) offers a straightforward way to manage risk attitudes in both individuals and groups, drawing on insights from the related fields of emotional intelligence and emotional literacy.

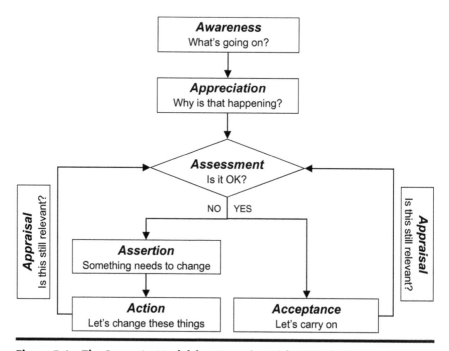

Figure 5.4   The Seven As Model for Managing Risk Attitude (From Murray-Webster and Hillson, 2021)

This model starts with *Awareness* since it is clearly impossible to actively manage something of which one is unaware. An individual seeking to manage their own risk attitude needs to be self-aware, able to diagnose their own current risk attitude. A degree of 'group self-awareness' is also required for groups wishing to deal proactively with the effect of risk attitude on group behaviour, either in managing risk or more generally in making decisions. It is also necessary for leaders in any position to be aware of the existing risk attitudes of those who they are leading ('others-aware'). Finally, Awareness needs to extend not only to the risk attitudes that are current, but to the effects these risk attitudes are having on the situation at hand.

The second A in the Seven A's model is *Appreciation*, leading to an understanding of why current risk attitudes have arisen or been adopted. This requires the ability to see the various triple strand factors at work, and to recognise where they have come from, regardless of whether that influence is valid and justified or not. It is necessary to understand the organisational context and culture, and how these might affect the situation and the people in it. Awareness identifies 'what' is going on, and Appreciation is about understanding 'why'.

Once the situation is understood through Awareness and Appreciation, it is necessary to decide what to do about it, if anything. This requires the third step in the Seven A's model, namely *Assessment*. In some circumstances, the unmanaged risk attitude adopted by individuals and groups may be fine, exerting no inappropriate influence on the situation. At other times it may be necessary to make a change if the existing risk attitude is leading to unhelpful risk management behaviour or poor decisions. For groups, Assessment is often undertaken by the leader in a given situation (for example, the project manager, project sponsor or risk champion), or by the group acting together, but any emotionally literate individual can adopt this role if they are aware of inappropriate risk attitudes influencing the group.

Following Assessment, the Seven A's model branches, depending on whether action is required or not. If the unmanaged risk attitude is OK and leading to appropriate behaviour and acceptable outcomes, the right response is *Acceptance*, allowing the current situation to continue. However, if something needs to be changed in order to allow a more appropriate risk attitude to be adopted, the next two As can be implemented, which are *Assertion* (creating the context for change through positive language and behaviour), and *Action*, when things are done to bring about the required change.

The Seven A's model also includes an *Appraisal* monitoring loop feeding back to Assessment, so that whether the decision is taken to intervene to change risk attitude through Assertion/Action or to leave it unchanged with Acceptance, the situation is regularly reviewed and reassessed to determine whether further change is required.

People Plus Process

This chapter has explored the importance of the human side of risk management and outlined how our perception of risk is a significant influence over the way we behave in risky situations, including both general decision-making and the various steps of the risk process.

Much more could be said than we have space for here, but the central point is clear: *people matter*. It is not enough to have a robust risk management process which is followed consistently. Every step in that process is performed by people, and each individual has a distinct personality, history, set of motivations and needs, relationships and so on. These characteristics influence how people react in the presence of uncertainty, both on their own and when in groups, leading them to adopt risk attitudes that vary from time to time and from situation to situation. This will have a significant effect on the risk process, influencing the risky decisions that we are required to make at each step.

Without taking proper account of the people aspects of managing risk, the risk process will be subject to unseen influences, leading to unreliable results and ineffective actions. Conversely, when the ways people respond to risk are fully understood and managed, then the risk process will work as it should.

Effective management of risk in projects (and elsewhere) requires both people and process, acting together to allow risk to be managed intelligently and appropriately. To deal properly with the people side of risk management we need to recognise the risk attitude spectrum and be able to place ourselves and other project stakeholders on it. This requires an appreciation of the triple strand influences, combined with a degree of emotional literacy that permits both understanding and modification of underlying risk attitudes, using the Seven A's framework to manage ourselves and others proactively. Only then can we execute the risk process properly, make appropriate risk-based decisions, and gain the full benefits offered by project risk management.

Chapter 6

Integrating Risk Management With Wider Project Management

In the years when risk management was developing as a discipline, it was perhaps natural for it to be treated as separate from mainstream project management. While people were unfamiliar with the concepts and practices of managing risk, it was necessary for there to be a distinct emphasis on risk management as a process in its own right, with its own particular set of tools and techniques, to ensure that it was properly understood and practised. Unfortunately, this initial separate focus has persisted beyond the early period when risk management in projects was becoming established, and it is still common to find organisations and projects where risk management is treated as an optional extra, additional to the core processes of managing projects, to be undertaken only for major projects (if at all), or if explicitly required by a particularly demanding contract or client.

This separation between risk management and project management leads to a loss of efficiency and effectiveness, and can prevent the risk process from making its proper contribution to project success. The goal should be for risk management to be 'built-in not bolt-on', becoming an integral part of the way projects are managed, rather than something to be done only under special circumstances. This chapter discusses the ways in which risk management can and should be integrated with the wider project management process. This integration is evident on two levels: firstly, ensuring that risk

DOI: 10.4324/9781003431954-6

management is built into the project lifecycle; and secondly, making clear connections between the outputs of the risk process and other project management processes.

Managing Risk Throughout the Project Lifecycle

We have seen in Chapter 2 that all projects are risky for three reasons:

1 As a result of their intrinsic nature;
2 By the deliberate design of projects as risk-taking ventures;
3 As a result of the environment and context within which projects are undertaken.

It is hardly surprising therefore that risk management should be an integral part of the way projects are managed. It is important however to know at which points in the project lifecycle risk management is relevant.

Discussing the applicability of risk management across the project lifecycle introduces an immediate problem. There is no universally accepted definition of a project lifecycle. Every project management standard or guideline seems to have a separate terminology, dividing the life of a project into a set of phases which differs from the others. Rather than arbitrarily choose one of these project lifecycles here, this chapter takes a more common-sense approach to mapping the contribution of risk management to the project lifecycle, which can easily be translated to the various lifecycle models currently in use. This discussion uses three simple stages to structure the way in which risk management is used across the project lifecycle, namely:

1 Before the project starts;
2 When the project starts;
3 After the project has started.

Before the Project Starts

The first question for any project lifecycle is 'When does the project start?' This question is simple to state but complex to answer. In fact, the lifecycle of a project is generally recognised as beginning before the project has started. A useful analogy is an individual human life, which most people would agree exists before the moment of birth. However, there is considerable controversy and ongoing debate about the exact time at which human life can be said to start, with strongly held competing views. This can be illustrated somewhat flippantly through various phrases that describe the extent of a human life. The term 'cradle to grave' is commonly used, indicating that a person exists from the moment of birth until they die. Two alternative and more light-hearted phrases give different perspectives on when human life starts however, speaking of 'womb to tomb' or 'lust to dust'. While the end point of a human life is reasonably clear (though there is some debate about this too), the moment at which life begins is less so. 'Cradle' suggests that life truly starts only when the process of birth is complete; 'womb' implies an earlier beginning at the time when a viable fertilised embryo becomes implanted; while 'lust' might refer to the moment of conception when someone has the desire to create a new life. Figure 6.1 illustrates these alternative viewpoints.

These analogies should not of course be over-interpreted, as we are only using them to reflect the range of views regarding when a project lifecycle can be said to commence. Some say that a project only exists after it is officially 'born', when there is a fully formed scope of work, with an agreed budget, schedule and completion criteria, and a project team in place to make it happen. Others contend that the project can be said to exist when it has become 'embedded' in the organisation as a viable entity, even if there might be some delay between this point and the time at which project work is started. Still others say that the project lifecycle starts when someone 'conceives' an idea that could give rise to a fully formed project, although many of these project

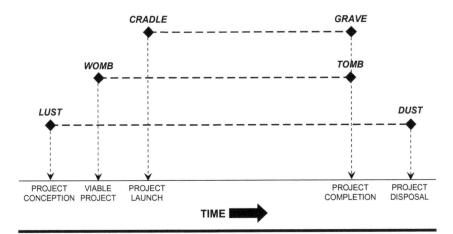

Figure 6.1 Alternative Views of Project Start and End-Points

concepts may be aborted and fail to be implemented. For the purposes of this chapter, we will adopt the latter definition, suggesting that the project lifecycle begins with a concept, which requires elaboration and development to determine whether it is feasible, prior to approval that the idea should be implemented as a project.

So how might risk management contribute to these pre-project stages of the lifecycle? Taking the three views outlined above, a clear understanding of risk exposure is important when someone initially articulates the desire to create a new project (conception), and also when determining whether a particular concept should be pursued (viability), as well as at the moment of birth when a project is actually launched (initiation). The role of risk management at these three points in the project lifecycle is as follows:

- *Conception.* Here we need to know the opportunities that a particular idea could present to the organisation, even though specific implementation details for any subsequent project might not be clear. These opportunities must result in clear benefits for the organisation and its stakeholders. It is also important to understand potential threats that the organisation might face in undertaking a project in this area. The risk process at this point should allow a risk-balanced decision to be made on whether to take the concept further, taking account of both upside and downside uncertainties, preferably using a risk-efficiency approach that balances risk and reward (discussed in Chapter 7). It should also encourage key stakeholders to determine their risk threshold for the idea, at least in high-level terms, to be used to inform later go/no-go decisions.

- *Viability.* Once the concept has passed the various organisational acceptability hurdles, including the test of risk-efficiency, and has been accepted as a potential project idea, its viability should be tested, to determine whether it is likely to survive and thrive as a fully formed project, and actually deliver the intended benefits. A range of alternative options for project implementation may be considered, and a relative risk assessment can be undertaken for each of these options to determine which is most likely to succeed. Applicable risk techniques might include decision trees, real options, Analytical Hierarchy Processing (AHP), influence diagrams or sensitivity modelling. The risk assessment might indicate that none of the identified options is feasible, with all alternatives being above the organisational risk threshold, in which case the project concept might be aborted, or further options could be developed. However, it is most likely that one option will emerge as the front-runner, with the best chance of delivering the intended benefits to the organisation.

- *Initiation.* If a feasible option is identified, the organisation may decide to launch a project to implement the concept, in full recognition of the level of risk inherent in undertaking the project. At this point a full project plan is developed, including a scope of work and Work Breakdown Structure (WBS), with realistic estimates of time, cost and resource requirements, and a project schedule is produced. These parameters are captured and documented in a project charter or business case, along with the associated underlying assumptions and constraints. A project manager and team are allocated to the project, and it is initiated as a fully formed project. Here the role of the risk process is to assess the risk inherent in the project which the organisation is proposing to undertake, including both the overall risk exposure of the project as well as key individual risks. This assessment is then compared with the organisational risk threshold to make a final decision on whether or not to proceed with the project, and what levels of contingency are appropriate if the project goes ahead. Understanding risk exposure at this stage will allow the business to determine how the project should be run after its initiation in order to control the inbuilt risk, while remaining flexible to respond to other risks that may emerge during the project.

It is important to recognise that the type of risk management undertaken during these pre-project stages does not strictly fall under the heading of 'project risk management'. Exploring the degree of risk exposure associated with a project concept and the feasibility of its various implementation options before deciding whether to pursue it actually belongs in the realm of portfolio (or programme) risk management. Nevertheless, when discussing the role of risk management through the project lifecycle, it is important to recognise that assessing and managing risk starts before the project itself commences. However, the major contribution of risk management to project success undoubtedly comes during the main part of the project lifecycle which occurs between project launch and project delivery.

When the Project Starts

The risk management process is of particular importance immediately after the moment when a project is approved and launched. At this point the organisation has committed to performing the project against a defined scope and requirement, with clear objectives and deliverables. However, although there may only be a brief elapsed time between the decision being made to initiate a project and its actual launch, there are often significant changes in this period, perhaps as a result of contract negotiations or internal project prioritisation processes. As a result, the risk exposure of the project actually being undertaken can differ significantly from what was expected in the last risk assessment of the project when it was authorised or approved for implementation. For this reason, it is useful for the project manager to undertake a full assessment of the risk exposure of the project as actually implemented, so that both the project manager and the team are fully aware of the overall level of risk exposure of the project which they are responsible for delivering. The results of this risk assessment can then be used by the project manager to determine the strategy for implementing the project.

After the Project Has Started

Although the risk management process has an important role in the pre-project phase and at the time of project launch, its greatest contribution to project success is probably during the project execution phase. This is the application area described in most project risk management standards and guidelines, which assume that a project already exists and has been launched, and then go on to describe how risk management should be undertaken during the rest of the project lifecycle. In this phase the risk process follows closely what has been outlined in Chapter 3, and it is not necessary to repeat it here.

One central question often asked when considering the use of risk management during project execution is how often the risk process should be performed during this phase. The answer is the typical response given by project managers to most questions of this type: 'it depends'. The frequency of application of risk management through the project execution phase depends on several distinct parameters, the most obvious of which is the degree of risk exposure in the project. Clearly a high-risk project is likely to require more frequent and detailed application of risk techniques than one which is lower risk. It is also possible that external clients or internal procedures may impose risk process requirements which the project has to meet, for example, providing risk updates at a frequency that matches the overall project reporting cycle (often monthly).

However, there is another reason why the risk process may need to be tailored during project execution. This relates to the project lifecycle approach being followed for the project. There are a number of different models of project execution in current use, of which the two most common are the sequential 'waterfall' lifecycle, and iterative project development models (also known as adaptive, spiral, agile or lean). The risk process needs to be applied differently for these alternative project development approaches.

The traditional risk process as described in most risk guidelines and textbooks applies mainly to the waterfall model of project execution. Here the project lifecycle is divided into distinct phases, each of which is completed before the project moves on to the next. It is common to have formal checkpoints at the end of each phase (often called 'gates') to ensure that the phase is indeed complete before progression to the next phase is approved. Within such a project lifecycle model, a complete iteration of the risk process is typically performed at the start of each project phase, to clarify the current risk exposure of the project at that point before the project team embarks on the next phase. The results of that risk assessment can then be used to determine the specific planning requirements of the next phase. If a given phase is particularly lengthy, it is common for the risk process to be repeated at interim points within the phase, to inform the decision-making of the project team.

In stark contrast to the linear and structured use of the risk process for waterfall project developments, risk assessment is used rather more creatively within iterative or adaptive project lifecycle approaches. The key characteristic of such development models is to divide the overall project functionality into a number of smaller elements (sometimes called 'chunks'), which are developed and delivered on a piecemeal basis, ultimately building up to deliver the whole project. It is a stated aim of iterative development that higher-risk elements should be developed and delivered early, in order to reduce the overall risk exposure of the project as a whole. The risk process is therefore used in this context to determine the relative risk exposure of each element, and to inform the sequence in which elements are scheduled for development. This should be done at the time when the project is first planned, but it should also be repeated at the end of each incremental delivery because the relative risk exposure of remaining elements is likely to have changed as a result of work undertaken on previous elements. The risk process also has a more traditional role within the development of larger incremental elements, to optimise the chances of successfully delivering each element. While this is clear in theory, the approach is rather less well developed in practice, and careful thought is required to ensure that assessment of risk is used appropriately when setting the agenda for incremental development project lifecycle models.

Contribution of Risk Management to Other Project Management Disciplines

The risk process naturally results in a better understanding of the areas of risk exposure on a particular project, and produces a set of targeted risk responses which aim to minimise threats and maximise opportunities in order to optimise the chances of the project achieving its objectives. However, the outputs from the risk process have much wider applicability than the obvious identification and management of individual risk events and overall project risk exposure. If the degree of risk faced by the project and the main specific risks that could impact the project are known and understood, this information can be used to shape and inform many of the other decisions and actions within the project. Elements of routine project management practice which can be enhanced by an understanding of risk include the following:

- *Contract negotiation and procurement (internal and external).* At the highest level a contract can be viewed as a vehicle for transferring risk between the contracting parties. In essence the vendor offers to perform some task for the buyer in return for an agreed consideration. However, risk information can be used at a more detailed level during contract negotiation and procurement activities. For example, it is possible to use the contract terms to specify which party will carry particular designated risks and what consideration will be given in return. Contracts can also be used to set up risk-sharing partnerships with clearly specified risk-reward conditions. It is also a good idea when conducting a competitive tender for the buyer to perform a relative risk assessment of the competing vendors, both in terms of their proposals and of their organisational characteristics. In this case the relative risk exposure associated with each bid will form one of the selection criteria in determining which bidder is successful. The same principles apply to internal procurement activities where formal contracts are not used, with risk information being used to clarify expectations and responsibilities between the parties, or to determine which procurement path to follow (for example, comparing in-house with outsourced).

■ *Baseline estimating (both time and cost).* Before a project is launched, estimates will be made of its expected duration and likely cost. These estimates are usually made on the basis of incomplete information and involve use of a number of assumptions, including scoping assumptions, planning assumptions, technical assumptions and so on. This introduces risk into the estimates, since the underlying assumptions used may be faulty, resulting in inaccuracies in the project schedule and cost estimates. A risk-based approach to estimating allows assumptions to be identified and challenged, and reveals the degree of uncertainty associated with project estimates. Simple three-point estimating (minimum, most-likely, maximum) for project time and cost should also be used to reflect risk, taking account of both estimating uncertainty and specific risks.

■ *Resource allocation.* One of the project manager's main tasks is to allocate appropriate resources to project tasks, often in consultation with line managers or functional managers. This is usually done by matching available skills to task requirements. However, it is preferable to adopt a risk-based approach to resource allocation where possible, by putting the more highly skilled people on the riskiest tasks.

■ *Selection of development options.* A project manager may wish to leave final decisions on implementation of some parts of the project until later in the project lifecycle, perhaps dependent on the results of earlier phases. Where more than one development option exists for the project, the relative risk exposure of each option should be assessed using a common framework, to enable the eventual decision to take proper account of risk, among other factors.

■ *Contingency management.* Contingency is used in projects at various levels, to cope with the effects of unforeseen risks. Different contingency funds may be allocated to the project sponsor, the project manager and project team members, to be applied under pre-specified conditions. The first challenge is to ensure that the right amount of contingency is allocated to the project, and this should obviously be a risk-based decision. Analysis of the overall risk exposure of the project should provide information on the range of possible project outcomes, allowing the organisation to decide how much contingency is appropriate at the various levels in order to give the required degree of confidence in project success. Risk information should also be used however to determine how and when contingency funds are spent, since they should only be applied if and when specific risks occur. Indeed, successful management of risk should allow the project to return unused contingency to the organisation in the form of additional margin or profit.

■ *Change control.* Most projects experience change at some point during their life, either imposed by clients and customers, or required by the organisation performing the project in response to changing circumstances. A formal change control process is adopted on many projects to ensure that proposed changes are assessed before being accepted, since changes in scope always result in changes in project time and cost. Of course, a proposed change may also change the overall risk exposure of the project, so risk assessment should form part of the change control process, to determine the degree to which risk exposure would be modified if the proposed change was accepted.

'Built-In, Not Bolt-On'

Project risk management is an important part of the management of projects, and makes a significant contribution to the chances of the project succeeding in meeting its objectives. As organisations focus on risk management to ensure that they are doing it well and gaining the expected benefits, there is a danger that risk management could be seen as somehow separate from wider project management. This chapter explains why such a view is mistaken, since it is impossible to divorce project risk management from its project context. Risk management plays a vital role at a number of key points throughout the project lifecycle, helping to ensure that the project is well specified, soundly launched and effectively executed. The outputs of the risk process should also be used to inform a number of other project management processes, providing a risk-based perspective which provides greater realism and robustness in these other processes.

While it may be necessary to focus on project risk management separately in order to ensure that it is performed properly, this should not be done at the expense of creating artificial barriers between project risk management and the projects it serves. Risk management will always provide some benefits to the project even if it is performed in isolation, by identifying key threats and opportunities and developing appropriate responses to deal with them in advance. However, the full benefits will only be attained by the project if risk management is fully integrated into the wider project context.

Chapter 7

The Bigger Risk Picture

In the previous chapter we saw how risk management in projects must not be treated as if it were separate from wider project management. Instead, it needs to be fully integrated into the way projects are managed if the management of project risk is to be fully effective and if the project is to gain the promised benefits. The phrase 'built-in not bolt-on' describes this well. There is however another level of integration which is important, which is addressed in this chapter.

Strategy, Tactics and Projects

Projects do not exist in isolation within an organisation. Properly understood, a project is part of the delivery mechanism for the overall strategic vision of the organisation. This is illustrated in Figure 7.1 (a simplification of Figure 2.1), which distinguishes strategy from tactics.

Organisations exist to create benefits for their stakeholders, and the corporate vision or mission statement defines the scope and extent of those benefits, as well as the change that is required to create them. This is shown in the left-hand side of Figure 7.1. However, vision alone does not create business benefits, and many organisations use projects as the change vehicle to deliver the capability which leads to the required benefits, perhaps managing related projects through higher-level programmes and portfolios (see right-hand side of Figure 7.1). Defining the desired vision, required change and

DOI: 10.4324/9781003431954-7

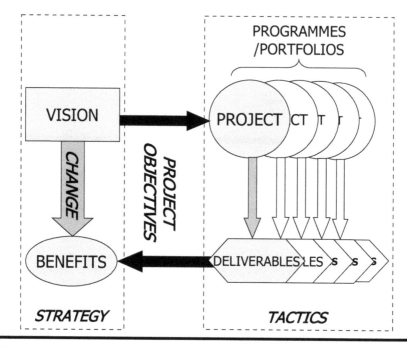

Figure 7.1   Strategy and Tactics

ultimate business benefits is the realm of strategy, whereas projects and their deliverables describe the tactics by which the strategy is achieved. Project objectives sit between the strategic and tactical levels, since they are defined in relation to the strategic vision, and they in turn define the requirement for projects (top arrow in Figure 7.1). Objectives are also used to measure the value of project deliverables (bottom arrow in Figure 7.1). Many projects fail because of a disconnect between strategic vision and tactical deliverables, often as a result of poorly defined project objectives. This space between the two levels of strategy and tactics requires careful and proactive management if projects are to succeed in delivering the required benefits to the business. Yet it is precisely in this area that businesses are most at risk.

Project objectives are affected by the uncertain environment within which projects and business are undertaken, resulting in a level of risk exposure. Project risk management exists to address this risk exposure, and should lead to an acceptable and manageable level of risk in each project. This increases the chance of meeting project objectives, which in turn maximises the likelihood of achieving the required business benefits. As a result, there is a clear link between project risk management and business performance: effective risk management at project level should lead to realised business benefits.

However, the project environment is not the only place where risk management is important, and successfully managing project risk is not the sole contributor to business success. We've seen that project objectives are (or should be) derived from the overall strategic vision of the organisation, but this is not typically done in a single step, except in very small organisations. More commonly, projects exist in the wider context of programmes and portfolios, creating a hierarchy of objectives within the organisation which progressively elaborates the strategic vision into more and more detail, eventually reaching the project level. Figure 7.2 depicts this hierarchy, showing how portfolios and programmes sit at intermediate levels between strategy and the resulting projects. (This figure is not intended to imply that these are the only objectives within a typical organisation, but merely to represent the range of objectives at different levels which lie on the path between the strategic vision and projects).

When deriving the business case for projects it is essential that there is a clear link with the strategic vision of the organisation, so that each project team understands how their work is contributing to achieving the wider purpose. This presents a double challenge to those responsible for management at every level in the organisation. The hierarchy of objectives produced through the planning process must exhibit both *coherence* and *alignment* if the tactical work is to deliver the strategic benefits.

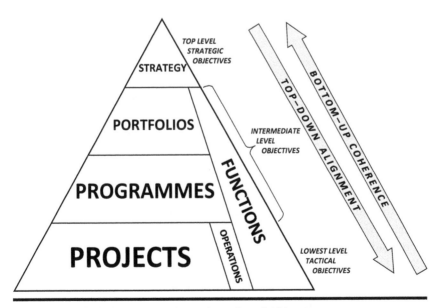

Figure 7.2   The Organisation as a Hierarchy of Objectives

Consequently, it must be possible to trace the overall vision down through the hierarchy, with *top-down alignment* as the vision is broken down into ever more detailed lower-level objectives. In the same way there should be *bottom-up coherence*, with the sum of the objectives on each lower level completely describing the next higher level. This demands attention to inter-level communication with the ability to both roll-up and drill-down through the hierarchy.

Hierarchy of Objectives, Hierarchy of Risk

In Chapter 1 we derived a working definition of risk as 'uncertainty that, if it occurs, will affect achievement of objectives'. Clearly in project risk management the focus is on finding and managing the uncertainties that could affect achievement of project objectives. But objectives exist elsewhere in the organisation, ideally as a coherent and aligned hierarchy. Wherever there are objectives, they are likely to be affected by uncertainty, whether that is at the highest strategic level of the organisation, through intermediate objectives, right down to tactical objectives within projects. In other words, risk exists at every level where objectives exist. And wherever risk is present, it should be managed proactively in order to maximise the likelihood of achieving the relevant objectives.

It is therefore possible to speak of different types of risk management, or more accurately, risk management with different levels of focus. So, one might use the term 'strategic risk management' to refer to management of strategic risk, which in turn can be defined as 'uncertainty that, if it occurs, will affect achievement of strategic objectives'. A range of similar specific definitions for various types of risk can be produced, describing financial risk, environmental risk, safety risk, operational risk, programme risk, and so on. Just as there is (or should be) a hierarchy of objectives across the organisation, so risk management is (or should be) hierarchical in nature. And in the same way that organisational objectives need to be coherent and aligned across the different levels, the management of risk at the various levels should be conducted in a coordinated manner.

It would of course be possible to manage risk at each level independently and without reference to the other levels. Indeed, this is currently the practice in many organisations, where there is little or no communication or linkage between different types of risk process. In these organisations there is no way of connecting management of financial risk by the treasury function to business-as-usual risks at operational level, or using project risk exposure to inform corporate governance decisions, or understanding how attempts to generate portfolio risk efficiency might influence reputational risk. The most effective way to address risk management within an organisation would be to have an approach that was integrated across the hierarchy. This is the realm of Enterprise Risk Management or ERM (also known as Enterprise-Wide Risk Management or Integrated Risk Management). There are however a number of ways in which ERM is understood, and it is important to be clear about the way we are using it here.

There are four distinct alternatives for how the term ERM may be used, and each of these is current in different settings. These four definition options for ERM relate to the type of risk that is considered to be in scope for ERM, summarised as follows:

1 *Risk 'to' the enterprise.* Here the goal for ERM is to manage any uncertainty that has the potential to affect the organisation as a whole. The word 'enterprise' is taken to refer to the entity, so ERM is management of risks to the entity. This leads to an exclusively high-level strategic and corporate view of ERM, and it is the clear responsibility of the organisation's senior leadership to implement risk management at this level.

2 *Risk 'of' the enterprise.* With this interpretation the scope of ERM is seen as being limited to the risk that the enterprise poses to its stakeholders. As for the first definition, this view of ERM leads to a high-level implementation, focused only on those uncertainties that could directly influence the major stakeholders. It tends to address the uppermost layer of risks, plus any which have filtered up from below because they have the potential for strategic impact.

3 *Risk 'within' the enterprise.* This perspective sees 'enterprise risk' as being the total of all risk exposure at all levels within the organisation, summed up to some aggregate amount. Again, management of this type of risk requires action at senior levels within the organisation, with the leadership attempting to use ERM to address the overall risk exposure of the whole enterprise.

4 *Risk 'across' the enterprise.* A fourth option is for the term ERM to be used to describe an integrated and hierarchical approach to management of risk across the enterprise at every level, including all those

shown in Figure 7.2. Here ERM has wide applicability at all levels and describes the combined effort of all staff to manage the risks that they face at their level, but to do this in an integrated way.

Although each of these variants of ERM is used by some organisations and risk practitioners, the last is the most widely accepted. Here we follow the fourth option, defining ERM as '*a comprehensive and integrated framework for managing risk at all levels in the organisation in order to maximise value*'.

Managing Enterprise Risk Across Boundaries

We have seen that a generic risk process can be applied at any level within an organisation in order to address any type of risk. However, this multi-level approach to risk management is not sufficient to ensure that the overall management of risk will be effective across the enterprise. This requires risks to be communicated between the various levels, to ensure that they are addressed at the most appropriate level.

In principle there are three potential sources of risk that could affect any particular level of an organisation. Risks can arise from three directions, as illustrated in Figure 7.3, namely *up* from lower levels (by *escalation* or *aggregation*), *down* from the levels above (via *delegation*), or *sideways* at the particular level itself. Each of these three routes are outlined in the following paragraphs.

1 *Risks from below.* There are two main ways in which any intermediate level can be affected by risks from lower levels.
   a) *Escalation.* Some individual lower-level risks are so large that they can affect the achievement of higher-level objectives. The word 'large' needs to be defined of course, since not all lower-level risks are relevant at the higher level. Escalation criteria are therefore required which will define the thresholds at which a lower-level risk should be passed up to the next level. These criteria need to include risks which impact higher-level objectives, as well as risks requiring responses or action at the higher level.
   b) *Aggregation.* It is also necessary to be able to aggregate lower-level risks, where a number of similar and related risks might combine to

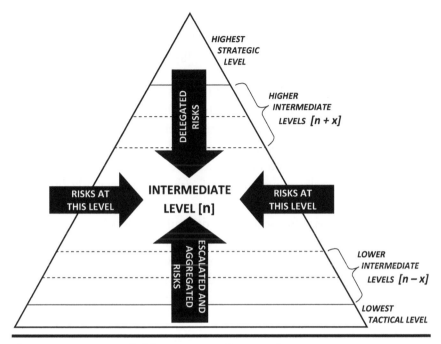

Figure 7.3   Sources of Risks at Intermediate Levels

create a higher-level risk, either by simple summation (ten insignificant lower-level risks may equal one significant higher-level risk), or as a result of synergy (the whole may be greater than the sum of the parts). Suitable risk categorisation schemes are required to facilitate such aggregation by identifying commonalities and possible synergies, and a generic Risk Breakdown Structure (RBS) may be used for this purpose (see Table 3.4 for an example RBS). When a group of aggregated risks becomes sufficiently significant, it is escalated in the same way as an individual large risk.

2 *Risks from above.* Organisational strategy is delivered by decomposing strategic objectives into lower-level objectives, as shown in Figure 7.2, creating an essential link between intermediate levels and those above them. There are strategic risks associated with the overall direction of the organisation, and many of these can and should be addressed wholly by the senior leadership of the organisation. However, some strategic risks will have implications for those lower levels which are used to deliver the strategy and create the business benefits. Strategic risks which can affect lower-level objectives or which require action at a lower level will need to be delegated. This requires well-defined

delegation criteria and thresholds, as well as clear channels of communication to ensure that management of strategic risks delegated to lower
levels is reported back to senior management. The goal is to achieve
delegation without abdication.

3 *Risks at the same level.* In addition to risks escalated and aggregated from
below or delegated from above, each level in the organisation is affected
by specific uncertainties arising at that level. These include both threats
and opportunities across the full range of risk types, including technical, management, commercial and external risks. These need to be
managed by application of the generic risk process previously discussed.

An essential prerequisite for escalation or delegation of risks is clear definition
of the boundaries and thresholds between the different levels in the organisation. This is necessary so that everyone knows where each risk belongs,
without confusion or ambiguity. Regardless of where a risk is identified, it
needs to be managed at the right level, and this is defined by the level of
the objective(s) which would be affected if the risk occurred. Quantitative
measurable escalation/delegation boundaries or thresholds can be produced
for each objective, against which a risk can be compared. Examples might
include the following:

- Total maximum (worst-case) impact of a single risk, not weighted by
  probability;
- Total aggregated unweighted (worst-case) impact of a group of related
  risks;
- Impact of a particular risk on corporate reputation;
- Impact of a single risk on one or more higher-level objectives.

A particular organisation may wish to add to these characteristics, depending
on the type of business or the nature of the risks faced. For example, some
organisations may wish to define thresholds based on safety or regulatory
compliance or share price etc.

Of course, whenever a risk is escalated or delegated between levels, this
must be followed by allocation of an agreed risk owner at the new level for
the risk, who accepts accountability for management of the risk. Effective
handover of escalated and delegated risks is essential to ensure that such risks
are managed properly, and that a risk does not fall between the levels with
each assuming that the other will handle it.

This generic scheme for escalation and delegation of risks to the appropriate level for management can be applied at any intermediate level in the organisation. There are however two places in the organisational hierarchy where the situation is slightly different, namely the top and the bottom. Referring back to Figure 7.3, it is evident that risks cannot be escalated to the lowest tactical, operation or delivery level, including projects. Similarly, delegation of risks is not relevant at the highest strategic level. This is where Enterprise Risk Management (ERM) comes in (Hillson, 2011).

Effective Enterprise Risk Management

Some view ERM as an unnecessary complexity, suggesting that the only requirement is to manage risk effectively at each level. They argue that if risk is dealt with at its point of origin wherever it arises within the organisation, then there is no need for an integrated approach that overlays additional bureaucracy. However, just as there are clear benefits to managing an organisation's objectives in a coherent and aligned manner, the same is true for managing risk.

ERM addresses risks across a variety of levels in the organisation, from strategic to tactical levels, and covering both opportunity and threat. Effective implementation of ERM can produce a number of benefits to the organisation at both project and higher levels which are not available from a non-integrated risk process. These include:

- Bridging the strategy/tactics gap to ensure that project delivery is tied to organisational needs and vision.
- Focusing projects on the benefits that they exist to support, rather than simply on producing a set of project deliverables.
- Identifying risks at the strategic level which could have a significant effect on the overall organisation, and enabling these to be managed proactively.
- Providing useful information to decision makers when the environment is uncertain, to support the best possible decisions at all levels.
- Creating space to manage uncertainty in advance, with planned responses to known risks, increasing both efficiency and effectiveness, and reducing waste and stress.

- Minimising threats and maximising opportunities, and so increasing the likelihood of achieving objectives at all levels from strategic to tactical.
- Allowing an appropriate level of risk to be taken intelligently by the organisation and its projects, with full awareness of the degree of uncertainty and its potential effects on objectives, opening the way to achieving the increased rewards which are associated with safe risk-taking.
- Development of a risk-mature culture within the organisation, recognising that risk exists in all levels of the enterprise, but that risk can and should be managed proactively in order to deliver benefits.

The good news is that ERM does not have to impose additional complexity or bureaucracy, if it is properly understood as integrated management of risk across the hierarchy. The basic risk management process outlined in Chapter 3 can be applied to the management of risk at any level, with a few simple modifications:

- The process is focused on achievement of the specific objectives at the level under consideration (for example, strategic risk management addresses uncertainties with the potential to affect strategic objectives).
- Risk-related tasks are performed by different people, namely those responsible for the specific objectives which are at risk (so strategic risk management is undertaken by senior management).
- Risk reports use the language of the stakeholders (for example, strategic risk reports relate to business benefits, share value, market position and so on).

The goal of ERM is to create an integrated approach to managing risk across all levels, with a shared understanding of risk by everyone involved, a common language for risk, the same risk process employed at each level, generic risk templates which are applicable for all, and a risk-aware culture across the organisation which recognises the value of risk management and is committed to implementing it effectively. One of the main success factors in getting this to work is an understanding of the boundary conditions and interfaces between the different levels of risk, to answer questions such as 'When does a project risk become a programme risk?' or 'How do strategic risks impact other parts of the organisation?' An effective approach to ERM will define such escalation and delegation criteria in terms of objectives at each level, ensuring that everyone has a shared understanding of how risk at their level relates to other levels.

Project Risk Management in the Wider Risk Context

Projects sit near the bottom of the hierarchy of objectives, connected to organisational strategy through several intermediate layers. As explained earlier in this chapter, it is clearly important for projects to be tightly coupled to strategic objectives, so that successful completion of each project and generation of its deliverables will make a positive contribution to creating value for the organisation and its stakeholders. In the same way, effective management of project risk is essential to achieving overall business benefits. In order to make this contribution, project risk management must have a clear working interface with higher levels in the hierarchy.

Management of risk at programme and portfolio levels has received some attention in recent years (for example Project Management Institute, 2019), but much of this guidance focuses on applying routine risk management processes at higher levels. However, managing risk in programmes and portfolios poses an additional challenge, which begins when they are first conceived and continue throughout their lifecycle.

Managing Risk in Programmes and Portfolios

As we consider how to manage risk in programmes and portfolios, an obvious question is whether we can simply take the standard project risk management approach and apply the same process at a higher level. But programmes and portfolios are different from projects, so risk management at these levels is different from project risk management. It is of course necessary to manage risks that could affect successful execution of a programme or portfolio once it is launched. However, risk management for programmes and portfolios is also important when the programme or portfolio is initiated for the first time, and when an existing programme or portfolio is reviewed at key points during its lifecycle. Here the challenge is to build and maintain *risk efficiency* (Hillson, 2023a).

This concept was first described by Markowitz when he introduced modern portfolio theory for financial investments (Markowitz, 1959). Risk efficiency involves developing various alternative versions of a programme or portfolio, each containing a different mix of candidate components. For each alternative version, the total overall risk exposure is determined, as well as the total expected return/benefit. The goal is to build or re-shape a programme or portfolio to ensure that an appropriate balance of risk against benefits is maintained.

The following steps are involved:

- *Understand risk thresholds.* It is important that the overall risk exposure of the programme or portfolio remains within the risk thresholds set by senior management, which in turn will reflect corporate risk appetite. This vital first step is often poorly understood or executed in many organisations. Without knowing how much risk is too much risk for the programme or portfolio, it will be impossible to know which components to include.
- *Assess potential components using a common risk/benefits framework.* The list of candidate components to be included should be well defined and understood, including the level of risk exposure associated with each one, and the extent of promised benefits. Both of these parameters need to be quantified using a common framework, so that they are directly comparable. This requires a good understanding of how to determine overall risk exposure for programmes, projects and other components.
- *Select components to balance risk exposure with benefits.* When each candidate component has been assessed for both risk exposure and potential benefits, the programme or portfolio can be constructed (if this is prior to launch) or re-shaped (at key review points during the lifecycle), using the principles of risk efficiency.

After generating alternative versions or the programme or portfolio, these are plotted on a *risk efficiency graph* that shows risk exposure against potential benefits. An example risk efficiency graph is shown in Figure 7.4. The *risk-efficient frontier* is shown on this graph, defining the optimal balance of risk exposure versus benefits, and separating options that have acceptable levels of either risk or benefit from other options. The region to the left of the risk-efficient frontier is infeasible, where the level of return cannot be achieved at the associated low levels of risk exposure. The region to the right of the

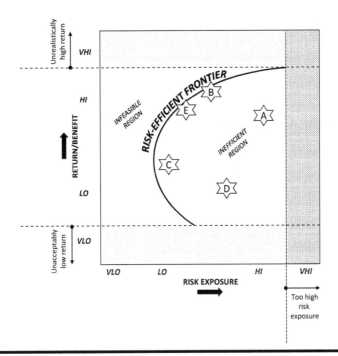

Figure 7.4 Example Risk Efficiency Graph (source: Hillson & Simon, 2020)

frontier contains feasible options, and those that lie closest to the frontier have optimal combinations of risk and benefit. Positions further to the right of the frontier are inefficient, where levels of promised benefits are too low to justify the associated level of risk.

In the example risk efficiency graph in Figure 7.4, five alternative versions of a programme or portfolio are plotted (A–E), each with different components. In this example, options A, C and D could all achieve higher returns for the given level of risk (i.e., they are inefficient), whereas options B and E lie on the risk-efficient frontier, with the maximum feasible return for the given level of risk.

Each option on the risk-efficient frontier should be considered as viable, but the actual version selected for execution will have a position on the frontier where the level of risk lies at or below the defined risk threshold and within the organisational risk appetite for the particular programme or portfolio. In Figure 7.4, if the risk threshold lies between options B and E, then E should be selected (lower risk than B).

Where review of an existing programme or portfolio shows that it has moved away from the risk-efficient frontier, due to either an unacceptably high level of risk exposure, or an infeasibly low level of achievable benefits, or both, then the mix of components should be reviewed and changed as necessary to restore risk efficiency.

Remaining Challenges

When managing risk in programmes and portfolios, we need to ensure that risk efficiency is achieved during their initial construction and that it is maintained throughout their lifecycle, as well as implementing a more traditional risk management process which addresses risks at each level which arise from a variety of sources. Although the principles described here are clear, at least three challenges exist in ensuring that they are implemented effectively:

- *Defining risk thresholds.* Effective management of programme and portfolio risk requires a clear understanding of how much risk is acceptable within a particular programme or portfolio. It is not possible to answer this question without understanding how to define corporate risk appetite and then express that risk appetite as quantified risk thresholds. Corporate risk thresholds must then be translated down to portfolio and programme levels. In addition, clear and agreed thresholds must be defined for escalation and delegation of risks between levels.
- *Implementing risk efficiency.* Building a risk-efficient portfolio using the techniques described in the preceding section requires an ability to turn the theory of risk efficiency into practice. Although the principles of risk efficiency have been known for over seventy years, and their application to financial portfolios is well established, it is not always immediately clear how such an approach should be applied at programme or portfolio levels. In particular, there may be challenges in deriving the measures required to build and operate a risk efficiency model, including both the ability to quantify the overall risk exposure of components on a consistent basis, and the ability to quantify promised benefits. Even when these two parameters are defined, many organisations are unsure how and where to set the risk-efficient frontier so that it reflects the risk appetite of key stakeholders for a particular programme or portfolio. Professional bodies recognise the concept of risk efficiency, but detailed guidance on implementation and application is still mostly missing.

■ *Avoiding a project mindset.* Many of the existing programme and portfolio management guidelines have been developed by organisations and practitioners who come from a project management background. As a result, their departure-point in terms of both thinking and process is the project mindset, and this is particularly true of their recommended approaches to managing risk. There are many pitfalls associated with taking a project-based view of programmes or portfolios, most of which arise from limitations in the underlying thinking. This is likely to remain a challenge for as long as the majority of programme and portfolio management practitioners continue to come from the world of projects, or at least until the concepts and practice of programme and portfolio management become better established.

Despite these challenges, there is no doubt that programmes and portfolios, like projects, are risky undertakings, and that successful delivery depends on effective risk management. Programme and portfolio risks can arise from above and below as well as within, and an effective risk management process must exist that can tackle all of these sources. In addition, risk management of programmes and portfolios must address levels of overall risk exposure through application of a risk-efficient approach to the construction and ongoing execution of the programme or portfolio.

All of this needs to be done in the context of a wider ERM framework if the organisation is to achieve its overall strategic objectives through execution of its constituent projects, programmes and portfolios. Merely addressing project risk is not enough, and a holistic and integrative approach is required.

Chapter 8

Risk Leadership

Risk is managed by people making decisions and choices and taking action. When seeking to manage risk in projects, most of those people have roles and responsibilities within the project arena. Indeed, in Chapter 3 we've already explored many of the risk-related tasks and activities undertaken by project personnel, including the project sponsor, project manager and project team members, the risk champion, risk owner and action owner, as well as other stakeholders (see Table 3.7).

However, successful management of risk in projects also requires others outside the project realm to play a part. These actors are not directly responsible for risk management in projects, but their attitudes and actions set an important context for project risk management. Their influence is less direct than the immediate project stakeholders listed in Table 3.7, but it is nonetheless vital. To distinguish them from people who are responsible for *risk management*, we refer to those with important indirect influence as exercising *risk leadership*.

The term risk leadership has been in use for some time (Hancock, 2010), but the concept has only recently received structured and detailed attention. For example, Barber describes it as embodying those elements of leadership that enable an organisation to deal well with uncertainty, and he provides detailed analysis of components of risk leadership in different settings (Barber, 2023). In this chapter, we explore the role of risk leaders in setting the context and environment within which project risk management can flourish and succeed.

DOI: 10.4324/9781003431954-8

Leadership Versus Management

Much has been written about the difference between leadership and management, not all of which is helpful. Commonly considered differences are summarised in Table 8.1.

Table 8.1 Leadership and Management (From Hillson, 2022, Used With Permission)

Leadership	*Management*
Leaders have followers	Managers have subordinates.
Leaders exert influence.	Managers issue instructions.
Leaders use motivational styles.	Managers use authority-based styles.
Leaders demonstrate referent power.	Managers use hierarchical, reward and sanction power.
Leaders define vision, strategy and desired outcomes.	Managers execute strategy through tactical action.
Leaders set boundaries.	Managers act within constraints.
Leaders define need for change.	Managers plan and implement change.

Several of these distinctions are relevant when considering the role of senior leaders and managers in the wider organisation in supporting management of uncertainty and risk, which is the realm of *risk leadership*, as distinct from *risk management*. Many of them are echoed in a range of high-level contrasts offered by Hancock to explain the term (Hancock, 2012):

- Guiding, rather than prescribing;
- Adapting, rather than formalising;
- Learning to live with complexity, rather than simplifying;
- Inclusion, rather than exclusion;
- Leading, rather than managing.

These general comparisons between leadership and management are however of limited value when seeking to understand what risk leaders actually do.

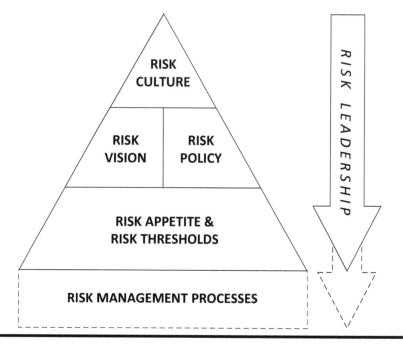

Figure 8.1   Risk Leadership Areas of Responsibility

There are four particular tasks which risk leaders are responsible for performing, as illustrated in Figure 8.1:

1. Shape risk culture;
2. Set risk vision and/or policy;
3. Define risk appetite and thresholds;
4. Influence risk management processes.

The four areas are presented in Figure 8.1 as a top-down hierarchy since each lower level flows from the one above. The most important aspect is the overarching *risk culture* of the organisation, which is expressed in a *risk vision* statement and/or top-level *risk policy*. Once the vision and policy are clear, it is possible to set the overall *risk appetite* of the organisation, and this in turn is expressed in a set of measurable *risk thresholds* that define acceptable degrees of variation against key strategic and corporate objectives. These risk thresholds can then be rolled out across the organisation in ever-increasing detail, informing personnel about how much risk is acceptable in their area of activity (and how much risk is too much). With a defined risk appetite and measurable risk thresholds, each level of the organisation can then implement their *risk management processes* within a clear framework that has been influenced from the very top of the business, ensuring that risk management across all levels is aligned with the risk culture and consistent with the risk vision and risk policy.

Before addressing these four areas in more detail, we need to consider where in the organisation risk leaders will be found.

Who Are Risk Leaders and What Do They Do?

The online Risk Leadership Network was launched in 2020 for risk leaders and their teams at major non-financial organisations, aiming to promote better practice by facilitating peer collaboration and knowledge-sharing (https://www.riskleadershipnetwork.com). This defines risk leaders as senior risk professionals with titles such as Chief Risk Officer (CRO) or Head of Risk, focusing largely on the Enterprise Risk Management (ERM) function. Many of the risk leadership characteristics outlined in the preceding section will be familiar to these senior risk specialists, who might already be demonstrating some or all of these attributes, and aspiring to others.

But risk leadership is not solely the preserve of risk specialists. Senior leaders in the business must ensure that their current style includes some risk leadership behaviours, expressed in partnership with expert risk colleagues who can provide the necessary support and input in the risk space at times of heightened uncertainty. Indeed, risk leadership roles and responsibilities are more likely to be performed by senior leaders in the wider organisation outside the risk specialisms.

Each of the four responsibilities shown in Figure 8.1 lies with the senior leadership of the organisation, and not primarily with risk specialists, though they will of course contribute, advise and support, particularly with the last.

In Figure 8.1, four distinct elements of risk leadership are presented, representing key responsibilities of those fulfilling the role of risk leader. Each of these four areas is explored in more detail in the following four sections.

Shape Risk Culture

Culture can be defined as 'the shared values, beliefs and knowledge of a group of people with a common purpose' (Hillson, 2013; Murray-Webster & Hillson, 2021). *Risk culture* is a subset of this more general phenomenon, describing 'the shared values, beliefs and knowledge *about risk...*', and explaining how the group views risk. This is driven by underlying *attitudes*

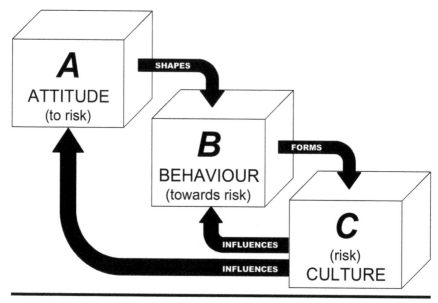

Figure 8.2   A-B-C Model (Adapted From Hillson, 2013)

towards risk, as well as the resultant outward and observed *behaviour* when risk is either encountered or perceived (Figure 8.2). Risk culture is exhibited by groups at different levels, including project teams, management review boards and the wider organisation within which the project is being performed. Without a clearly defined and articulated risk culture for the whole organisation, different areas will develop their own way of thinking about risk which may or may not be coherent and consistent with other areas. This leads to divergence on the way risk is handled within the organisation, with inefficiency, miscommunication, and room for avoidable errors and missteps.

Risk leaders are responsible for setting the bounds and parameters of risk culture. For senior leaders in the organisation, this means defining, articulating, communicating and modelling how they intend people in their business to think about risk, at all levels from top to bottom. The A-B-C Model in Figure 8.2 shows that risk *culture* is formed by repeated *behaviour* towards risk, which in turn is shaped by underlying *attitudes* towards risk. For risk leaders to influence risk culture, they first need to shape those attitudes held by people at all levels across the organisation.

These attitudes can be defined in terms of a 'risk mindset', which outlines in more detail the 'shared values, beliefs and knowledge about risk' contained within the risk culture. A positive and mature risk mindset includes the following six values:

- *Risk is natural.* Life itself is uncertain, and this translates into every type of human endeavour. The risk mindset accepts this reality and doesn't struggle to remove all risk from every situation.
- *Risk is manageable.* There is always something we can do in response to each risk. Avoid a victim mentality – we are not powerless, as long as we can see risk in advance. The risk mindset always seeks to influence risk, either by tackling its occurrence or addressing its effect.
- *Not all risk is bad.* Some uncertainties can be unwelcome if they occur, causing delay, damage or disruption. But other uncertainties would be helpful if they happened, resulting in reduced costs or timescales, or enhanced performance and reputation. The risk mindset remains alert to both threats and opportunities.
- *Risk matters.* Risk is always linked to objectives. If a risk happens, it will affect our ability to achieve one or more objectives, for better or for worse. Every risk is important, although some are more important than others. The risk mindset is relentlessly focused on objectives.
- *My risk is my responsibility.* It's easy to think that someone else will address risks, and it's 'not my job', especially at work. But where risk affects my objectives, I need to deal with it. The risk mindset takes responsibility for relevant risks.
- *Proactivity is essential.* We often adopt a 'wait-and-see' attitude to risk, hoping that we might 'get lucky', with bad things not happening and good things just turning up. The risk mindset rejects wishful thinking and unrealistic optimism, understanding that prompt action is often required.

It is easy to see how these six attitudes in the risk mindset translate into a particular set of risk behaviours, which will form the organisational risk culture. If one or more of these values were to be different, the approach to understanding and managing risk would be correspondingly altered. For example, believing that risk is not natural leads to an expectation that it should be possible to conduct our activities without risk, if only we could remove it all. But if we think that risk is essentially unmanageable, we become passive victims in the face of 'inevitable uncertainty'. This lack of proactivity is reinforced by the view that dealing with risk is the responsibility of risk specialists, and other staff do not need to do anything towards its management.

A completely different set of behaviours is encouraged when the six values listed above are firmly held and consistently expressed. We then view risk as a normal part of our business, but accept that it can and should be managed proactively, and that each member of staff needs to identify and manage risk in their own areas of responsibility. Understanding that all risk matters, and that some risk offers additional upside benefit leads us to take more risks in the pursuit of value, being proactive wherever possible to minimise threats and maximise opportunities as we do so.

Set Risk Vision and/or Policy

Once the values in the desired risk culture are defined and communicated, senior leaders in the organisation can move on to express these values in a formal risk vision statement or a top-level risk policy. Some businesses may prefer to use vision statements instead of policies, or vice versa, and some may be familiar with both. The important thing is to have a simple, clear declaration of intent for the whole organisation, from which a complete risk management approach can be developed, and against which it will be measured and tested.

An example high-level risk vision or policy statement might read as follows:

> Many aspects of our business involve risk, and we recognise that success depends on our ability to manage risk effectively in order both to protect value and to create value. We aim to make risk-informed decisions at all levels across the business, taking account of both threats and opportunities in pursuit of sustainable business benefit, competitive advantage and operational efficiency. Identifying and managing risk is an integral part of our strategy, decisions, operations and projects.
>
> Responsibility for managing risk effectively is spread across the business, starting with top management, and distributed across all staff. In particular, leaders at all levels are committed to identifying, understanding and responding to risk, and supporting their teams in its management.
>
> We foster a risk-aware culture in all decision-making, and are committed to managing risk in a proactive and effective manner. We support this with an integrated framework of risk governance and reporting, analysing risk in order to inform management decisions taken at all levels in the organisation, and

taking appropriate action to deal with those risks that exceed the thresholds that can be tolerated for our objectives.

Our Enterprise Risk Management framework supports consistent and effective management of risk across the business. Through balancing risk-taking within defined risk thresholds, we will generate increased value for our stakeholders, increasing their confidence in our ability to meet our strategic and business objectives.

Depending on the practice of the organisation, the risk vision/policy may be supported by a more detailed risk management standard or risk framework. Development of these documents may be delegated to senior risk professionals, but organisational leadership needs to retain oversight of their content, ensure that they faithfully reflect the agreed vision/policy, and approve or authorise their issue.

Define Risk Appetite and Thresholds

With the values that underpin risk culture in place, and a clear statement of risk vision and/or risk policy, it is now possible to move forward with setting the context for managing risk across the organisation. An essential first step is to answer the question 'How much risk is too much risk?' It is not possible to embark on risk management without knowing the target level of risk above which exposure is unacceptable. Many organisations go straight to the stage of setting risk thresholds in order to answer the question, but there is a prior step that must come first. This is understanding and defining risk appetite.

Risk appetite has received a lot of attention in the past decade or so, with some confusion over the precise meaning of the term. Comparison with natural appetite indicates that risk appetite is rightly understood as *the tendency of an individual or group to take risk in a given situation* (Murray-Webster & Hillson, 2021). The internal nature of risk appetite makes it hard to measure or quantify, and this is where risk thresholds come in, as measurable expressions of risk appetite set against specific objectives. Risk thresholds give a quantified measure of variation, stating upper and lower limits of acceptable uncertainty against each objective. Where risk appetite is internal and hard to influence, risk thresholds can be consciously modified to meet the risk challenge of specific situations, allowing the organisation to choose to take more or less risk in pursuit of their objectives (Murray-Webster & Hillson, 2021).

Risk leadership includes the responsibility of senior leaders in the organisation to understand their underlying risk appetite in respect of corporate or strategic objectives, and to translate this into quantified risk thresholds that can guide risk management activities. While this is rightly something for senior leaders to undertake, they are likely to need assistance from expert risk colleagues to facilitate the process.

Having set top-level risk thresholds against strategic objectives, it is then possible for leaders across the remainder of the organisation to define risk thresholds for objectives in their own area of responsibility, decomposing strategic risk thresholds to their component parts in an aligned way, so that the overall risk management effort of the organisation is coherent in remaining within the defined corporate risk appetite and agreed risk thresholds.

Influence Risk Management Processes

Risk leaders set the context for effective management of risk through the following actions:

- Promoting values that underlie a mature risk mindset that leads to a culture that is risk-aware and risk-mature;
- Embodying these values in a formal statement of risk vision and/or risk policy that can act as a touchstone for management of risk across the organisation; and
- Articulating the overall risk appetite for the organisation and expressing these in measurable quantified risk thresholds against the highest strategic objectives.

The final area where risk leadership is relevant is in implementation of the approach to managing risk. This of course is the domain of risk management rather than risk leadership. However, risk leaders have an important role here in influencing the way risk management is performed, even if the activities are undertaken by others.

There are a range of ways that risk leaders can influence the way others tackle managing risk. A combination of the following approaches should provide the necessary encouragement:

1 *Mandate it*. Where possible, risk leaders should insist that structured risk management processes are implemented in their area of responsibility. While it is not the whole answer, it sometimes helps to tell people

what to do. If a risk leader will always ask for risk information as part of routine governance, then the team will know that it matters.

2 *Simplify it*. Risk management need not be complicated. Risk leaders should encourage people to make the risk process as simple as possible without compromising effectiveness. The bureaucratic and administrative overhead should be minimised, keeping the process focused, and only collecting information that will be used.

3 *Normalise it*. Risk leaders must explain that managing risk is normal, and not an optional extra. Risk management needs to be built into every activity within the organisation.

4 *Demonstrate it*. A risk leader will lead by example and be a role model, identifying and managing risk in their own daily tasks.

5 *Use it*. Risk leaders will ensure that their strategic decisions take full account of risk information, showing people that their efforts make a difference to the wider organisation.

6 *Celebrate it*. Risk leaders will look for proof that risk management has tackled a threat so that a problem was avoided, or evidence that a potential opportunity has been converted into a real advantage. These successes should be communicated so that people know risk management works in practice.

These steps should ensure that people in the organisation know how importantly the risk leader views risk management, and encourage them to take it seriously and do it properly. They represent both 'push' factors (items 1–3 in the list above) that require people to manage risk in their own areas of responsibility, and 'pull' factors (items 4–6) that inspire or draw them towards doing the right thing. It may not be possible for a particular risk leader to employ all six of these sources of influence, but using a combination of multiple factors will increase the chances of success.

Behaviours of Effective Risk Leaders

The ability to successfully perform a risk leadership role requires more than simply performing a series of activities. There are a distinct set of behaviours that are frequently exhibited by successful risk leaders, relating to how they approach risky and uncertain situations. These include the following (Hillson, 2022):

- Risk leaders understand and welcome the uncertain nature of the future, with multiple possible outcomes.
- Risk leaders recognise both downside risks (threats) and upside risks (opportunities), and they are determined to take proactive action to minimise threats and maximise opportunities, balancing value protection with value creation.
- Risk leaders are prepared to invest resources in risk management in order to maximise the chances of positive outcomes, even when results are not immediately evident.
- Risk leaders can answer the question 'How much risk is too much risk?' for the business, by defining the limits of acceptable risk exposure as measurable risk thresholds that express risk appetite.
- Risk leaders encourage appropriate risk-taking within agreed limits.
- Risk leaders demonstrate personal commitment as risk owners by identifying and managing threats and opportunities in their area of responsibility, leading by example, and empowering others to take responsibility for managing risk in their own areas of responsibility.
- Risk leaders understand the strengths and weaknesses of the prevailing risk culture, and influence culture change towards a more mature set of beliefs and behaviours.
- Risk leaders recognise that outcomes will not always match plans, some risks will materialise despite people's best efforts, and blame is usually not appropriate.
- Risk leaders welcome those who make them aware of previously unseen risks (including whistle-blowers).
- Risk leaders learn from experience and take action to avoid repeating past mistakes.

Recent work has detailed specific risk leadership attributes and behaviours exhibited by those in executive roles as well as by managers and staff at lower levels in the organisation (Barber, 2023). This research was focused on the challenges faced by risk leaders in situations of high complexity, but much of the findings are applicable more widely. The same research also lists common behaviours exhibited by risk leaders at all levels, some of which are not only required in complex situations, including:

- Risk leaders never ignore or hide risks.
- Risk leaders network widely, involving stakeholders and others in seeking out new ideas and views.

- Risk leaders listen to and respect the views of others, without blaming or taking sides.
- Risk leaders seek out factors, positive or negative, that could influence success.
- Risk leaders proactively engage in risk processes.
- Risk leaders proactively collaborate with colleagues to help them understand and address risks.
- Risk leaders seek to maximise future success, not just meet targets.
- Risk leaders avoid oversimplification and untested assumptions.
- Risk leaders seek outcomes that optimise risk as a whole.
- Risk leaders never ignore or cover over difficult or sensitive issues or risks.
- Risk leaders show courage when required.

Risk leadership is also supported by several generic behavioural attributes that are demonstrated by effective leaders. These include (Strella et al., 2015):

- *Curiosity*. Risk leaders are driven to proactively seek understanding and new learning through gaining new ideas, experiences and information. This curiosity is expressed through welcoming change, experimentation and feedback.
- *Insight*. Risk leaders are able to process a vast range of information from many kinds of sources, using it to shape insights that make sense of ambiguity and simplify complexity. This requires a combination of conceptualisation skills, creativity and energy.
- *Engagement*. Risk leaders connect with people, resonating with the motivations and priorities of others. Their own enthusiasm, energy and sense of purpose are infectious, engaging others to deliver shared objectives and mutual benefits.
- *Determination*. Risk leaders enjoy a challenge, seeking to overcome obstacles. This enables them to take on higher-risk opportunities with tenacity, exercising the required self-discipline to focus and stay with the challenge, but without becoming stubborn or inflexible in the face of disconfirming evidence.

Risk Leaders – An Essential Contributor to Risk Management

Successful management of risk in projects depends on more than just what happens within the project itself. Projects are not conducted in isolation, and their context is a critical success factor. One key element of the project environment is whether senior managers and leaders in the wider organisation are providing the necessary support and encouragement to enable and facilitate project risk management, and this is where risk leaders come in.

There are other factors that are necessary if risk in projects is to be addressed in a way that is sustainable, and these are tackled in the final chapter.

Chapter 9

Sustainable Risk Management

The preceding chapters have laid out the case for risk management in projects, starting from the key underlying concepts that link uncertainty to the nature of projects and their objectives. A generic project risk management process has been outlined, and special considerations applying to complex projects have been described. The importance of understanding how people respond to risk when they have to make risk-based decisions has been discussed. Finally, we have considered the wider picture, relating management of risk to project management and to higher levels within the organisation, and explaining the role of risk leaders across the organisation beyond the project management area. While it is undoubtedly true that risk management is not hard, there are a number of challenges which those wanting to manage risk in their projects need to address and overcome. The earlier chapters of this book should have persuaded the reader that risk management is an essential component of how projects should be managed. But anyone who takes a casual approach to managing project risk is likely to encounter difficulties. For many people these arise from an unthinking and simplistic reliance on process to manage risk, without taking account of the people aspects. Even those who are aware of the influence of human nature on the risk process may find problems if they treat project risk management in isolation from its project management context or the broader organisational setting.

In order for project risk management to deliver its promised benefits, there are a number of Critical Success Factors (CSFs) which must be in place,

DOI: 10.4324/9781003431954-9

and some of these are discussed in this final chapter. CSFs have two characteristics which make them 'critical':

1 It is not possible to succeed in their absence.
2 If they are present, the chances of success are maximised.

The role of CSFs in supporting the risk process can be explored through a novel approach known as 'risk energetics', which is described in this chapter.

Risk Energetics

The dashed line in Figure 9.1 illustrates the natural decay curve which is experienced by an energy pulse in a free and unconstrained setting. A rise in energy follows the initial input, but this quickly starts to decline and ultimately reaches zero. This decay curve can also be used to illustrate the level of energy that is evident in a group of people who are seeking to manage risk (for example, a project team), if their situation is unmanaged and without external input. Following a period of initial enthusiasm, their degree of engagement soon peaks and starts to reduce, until they eventually lose interest in the risk management process. This may be due to natural busyness and tiredness resulting from the day-to-day work of performing the project, or

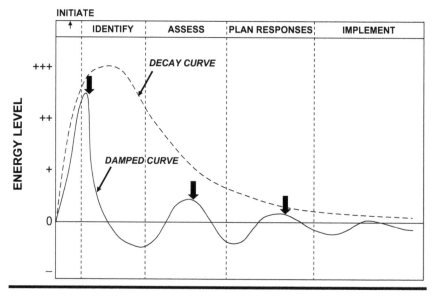

Figure 9.1 Risk Energetics – Decay and Damped Curves

may arise from other distractions that prevent the team applying themselves to the risk process. However, some project teams experience active discouragement and barriers which can lead to a damped curve as shown by the solid line in Figure 9.1, resulting in negative energy and failure to engage at all with the risk process. The reasons people give for not paying proper attention to managing risk in projects are explored in more detail elsewhere (Hillson & Simon, 2020).

If the steps in the risk process are overlaid onto this energy decay curve as shown in Figure 9.1, the natural unmanaged progression of a group of people undertaking risk management can be illustrated. This indicates initial enthusiasm when the risk process is first launched, peaking during the risk identification step. The peak probably occurs because this step is interesting and engaging, giving the team the chance to raise their concerns about risks on their project, and allowing their worries to be documented as threats that could negatively affect the project, while also capturing good ideas as opportunities that might assist the project. The use of creative techniques such as brainstorming or workshops also generates a sense of excitement, leading to raised energy levels.

From this point on however, the level of energy in the team tends to decrease with time. There is less enthusiasm for the risk assessment task, which can be seen as a chore, having to discuss each of the identified risks and consider their probability of occurrence, degree of impact, ownership, proximity, urgency, and so on. The energy level reduces still further when the risk response planning step is reached, leading to a tendency for teams to take the first feasible response instead of taking care to examine alternatives and select the most effective option. Finally, the unmanaged energy curve gets close to zero in the most important step of the risk process, when agreed risk responses are actually implemented. At this point the project team are likely to have lost interest in the risk process, perhaps even viewing it as a distraction from their 'real project work'. Any risk responses allocated to them may not get the degree of attention they deserve, and implementation may be cursory or superficial (or absent).

Obviously, this situation is not likely to lead to effective management of risk on projects. As a result, active intervention is required in order to ensure that energy is maintained at a sufficiently high level to promote and support an effective risk process. This intervention can have two aims: to reduce the effect of influences that dampen the energy curve to produce a decay, or to stimulate additional energy and maintain the required high level. The desired energy level is shown in Figure 9.2 (solid line), overlaid above the unmanaged

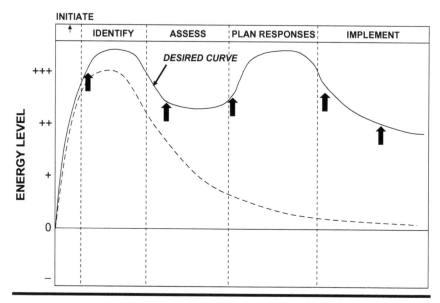

Figure 9.2 Risk Energetics – Desired Curve

decay curve for comparison (dashed line). In this curve, interventions are made to keep energy levels up, particularly in the two most creative phases of the risk process, namely risk identification and risk response planning.

Active inputs to prevent decay and maintain energy during the risk process can be viewed as CSFs. As already mentioned, CSFs have two characteristics: their presence promotes effectiveness, and their absence hinders it. Similarly, intervening actively in the risk process to maintain energy levels will contribute to a more effective process, and failure to intervene will result in reduced energy and process ineffectiveness.

Some of the more significant factors which affect the risk process energy level are described in the following sections, divided into two groups. The first of these are internal factors that are within the scope of the project itself, and which can probably be implemented directly by the project team. The second group of factors are external to the project and are the responsibility of the wider organisation to provide, including risk leaders.

Internal Factors

Three particular groups of internal factors deserve mention here, though there are others.

Process Design

One of the dampening influences over the risk process which can quickly sap energy and enthusiasm from the team is the design of the risk process itself. Where the process is bureaucratic or complex, people will soon disengage from it. This barrier can be overcome by thoughtful *process design,* seeking to maximise efficiency and reduce the overhead associated with running the risk process, while not cutting any essential corners. Use of templates can also assist in reducing the burden of data capture and recording.

It can be helpful to introduce a *process break* to reduce energy loss. For example, it is common to use a risk workshop setting for the identification and assessment stages, and sometimes these workshops are extended to include preliminary risk response planning. Since both risk identification and risk response planning require use of creativity and original thinking, it is asking a lot of project teams to expect them to maintain a high level of engagement and interest for a long time in a workshop. Instead, the workshop could be split into two or three elements, covering risk identification in the first, followed by a break, then going on to assessment and possibly also response planning at a second session. Sometimes it is enough simply to take a lunch break in the workshop, identifying risks in the morning and assessing them in the afternoon. Alternatively, a two-day workshop can be arranged, ensuring that participants have the chance to recharge their batteries and come fresh to the second instalment.

Facilitation

A proven contributor to maximising risk process efficiency is the use of a skilled and experienced *facilitator* (Murray-Webster & Pullan, 2023). This person can have various titles, such as Risk Champion, Risk Coordinator, Risk Process Facilitator, or Risk Manager. More important than their job title however are their personal characteristics. A good Risk Champion will have a combination of *technical skills* (including both the domain of the project as well as technical risk competences) and *people skills* (including the ability to understand and manage different types of individuals and groups). These latter soft skills are very useful for keeping energy levels high during the risk process, and a high degree of emotional literacy can be particularly helpful.

Where a Risk Champion is used to facilitate the risk process for a particular project, they should take responsibility for its effective and efficient operation. This is likely to include briefing the team on the purpose of risk management, leading workshops, recording outputs, drafting reports and chasing progress on actions. The ability to encourage and motivate people in these settings is key to a successful risk process, and will ensure that project team members stay engaged and enthusiastic about managing risk on their project.

It should be noted that 'Risk Champion' is a role and may not necessarily equate to a single individual on every project. Some organisations may indeed allocate a dedicated Risk Champion to each project, at least for major or large projects. Others may provide part-time Risk Champions from a central pool outside the projects, perhaps via a Project Management Office or Risk Competence Centre. Another alternative is for the Risk Champion's duties to be undertaken part-time by another team member, perhaps even the project manager. It is more important that someone facilitates the risk process than where they come from in the organisation.

Resources

It is evident that risk management is not a cost-free activity, and the project needs to provide the necessary level of *resources* if the risk process is to function properly. These resources include *people, time* and *money*. Of these three, people are undoubtedly the most important, and the project should ensure that the team includes members with the necessary experience and skills to undertake effective risk management (some organisations use the acronym SQEP to indicate the need for Suitably Qualified and Experienced Personnel). However, the risk process cannot succeed if it is not allocated adequate time, and the project schedule should explicitly include risk-related tasks such as risk workshops, risk reviews and so on. Similarly, an amount must be included in the project budget for both the risk process and for the cost of implementing agreed risk responses.

External Factors

In addition to factors that are under the control of the project itself, there are a range of external influences that contribute to the overall effectiveness of the project risk management process. These can also be grouped under three headings.

Infrastructure

The organisation is responsible for ensuring that each project has the necessary infrastructure to support the various activities and processes of the project. This is usually provided as a generic organisational capability into which each individual project taps.

We have already seen that although there is a core risk process to be followed, the level of detail required can vary from one project to another. Low-risk projects may only need a simple risk process, whereas more challenging projects might require a more in-depth approach. In the same way, different organisations may choose to implement risk management in varying levels of detail, depending on the type of risk challenge they face. The decision over implementation level may also be driven by organisational risk appetite, and by the availability of funds, resources and expertise to invest in risk management. Each organisation must determine a level of risk management implementation which is appropriate, acceptable and affordable. Having chosen this level, the organisation then needs to provide the necessary infrastructure to support it.

At its most simple, risk management can be implemented as an informal process in which all the phases are undertaken with a very light touch. At the other extreme is a fully detailed risk process that uses a wide range of tools and techniques to support the various phases. The typical organisation will probably implement a level of risk management somewhere in between these two.

Having selected the level of implementation, the organisation must then provide the required level of infrastructure to support the risk process. This might include choosing techniques, buying or developing software tools, allocating resources, providing training in both knowledge and skills, developing procedures which integrate with other business and project processes, producing templates for various elements of the risk process, and considering the need for support from external specialists. The decision on the required level for each of these factors will be different depending on the chosen implementation level.

Failure to provide an appropriate level of infrastructure can cripple risk management in an organisation. Too little support makes it difficult to implement the risk process efficiently, while too much infrastructure adds to the cost overhead and presents bureaucratic barriers. Getting the support infrastructure right is therefore a critical success factor for effective risk management, enabling the chosen level of risk process to deliver the expected benefits to the organisation and its projects.

Organisational Risk Culture

The importance of risk culture has been explored in Chapter 8, focusing on the role of senior leaders in the organisation to define, shape and influence the way their organisation thinks about risk. Organisational risk culture is a major topic which presents a multi-dimensional challenge to the business which is serious about managing risk effectively. Here we will concentrate on those elements of organisational risk culture that contribute towards effective risk management. Perhaps the most important of these is a culture which is *risk-aware*, recognising the existence of risk both within the business and in the external environment, as well as intrinsically present in the projects being undertaken by the organisation. Denial of risk is fatal to the ability of an organisation or its projects to manage risk properly, and conversely acceptance of its existence is a prerequisite to its management.

A second characteristic of appropriate organisational risk culture is to be *risk-mature*. This describes a culture which has a well-developed approach to risk at all levels, which is not surprised when risk is encountered, and which is able to take risk in its stride. A risk-mature organisation takes a proactive approach to risk management in all aspects of the business, makes active use of risk information to improve business processes and gain competitive advantage, and learns from its experience.

A last element of risk culture that has a significant influence on whether the project risk process is effective or not is the way risk-taking is regarded. The organisation (and particularly its senior management) should *encourage and reward appropriate risk-taking,* and will celebrate successes when projects and their teams demonstrate an effective approach to managing risk. Where the converse occurs and people are punished or discouraged from taking any level of risk, this will result in a lack of commitment and enthusiasm for the risk process and reduced effectiveness.

Management Support

The role of management in encouraging and rewarding appropriate risk-taking has already been mentioned, but there are other things that senior managers can do to maximise the effectiveness of the risk process on their projects. These revolve around demonstrating a *visible and consistent commitment to risk management,* with two particular aspects.

The first way senior management can show their commitment to the risk process is to appoint a senior manager (who may be called the Corporate Risk Sponsor or similar) who will promote the cause of risk management at the highest levels of the organisation. This role is ideally filled by a Board member, responsible to the CEO and the Board for setting risk policy for the entire organisation, creating a 'pull' for risk management from the lower levels of the business. The Corporate Risk Sponsor is also responsible for receiving risk reports from within the organisation on behalf of the Board, and ensuring that their content is complete and correct. The Corporate Risk Sponsor is effectively the 'end-user' or 'customer' for risk information produced by the business, and acts on behalf of the CEO and Board.

The Corporate Risk Sponsor may be supported by another senior role, perhaps called the Corporate Risk Champion, who has a central coordinating role within the business, acting as a focal point for implementation of all types of risk management activities at all levels across the organisation. The Corporate Risk Champion acts as the 'advocate' of risk management activities, and is responsible to the Corporate Risk Sponsor for setting performance criteria for risk management implementation, providing expert guidance at all levels, and supplying assurance to the business that lower-level risk processes are functioning effectively in compliance with the overall risk policy set by the Corporate Risk Sponsor.

The second major way in which the senior management of the organisation can demonstrate commitment to effective risk management across their projects is to use the results of the risk process to support *risk-informed decision-making*. When project teams can see that their risk information is actually being used to assist senior managers in running the wider business, they will be motivated to provide the best possible outputs from the project risk process. Conversely if the risk process is confined to the project level and its results are never seen by senior management, or worse, they are seen but ignored, project teams will quickly learn that there is no point in investing energy in managing project risk.

Risk Energetics Across the Project Lifecycle and Beyond

Figure 9.1 suggests that project teams engaged in an unmanaged risk process will inevitably lose energy and enthusiasm as the risk process progresses, and active discouragement will hasten and deepen the rate of decay. There are however a wide range of factors that can be deployed to counter the natural loss of energy, leading to a consistently higher level of energy throughout the risk process (Figure 9.2). These two figures illustrate the position across a single iteration of the risk process from risk process initiation to risk response implementation. However, we have learned in Chapter 3 that risk management is not a single-shot process, but it should continue during the project with a series of risk reviews, to ensure that the project remains aware of its current risk exposure and responds appropriately. This is reflected in Figure 9.3, where the risk energetics cycle is extended into a series of risk reviews and subsequent implementation of newly identified risk responses. The figure shows that renewed input of energy is required at the start of each update cycle in order to maintain the effectiveness of the risk process throughout the project lifecycle.

Of course, Figure 9.3 only describes the position for a single project, and one would naturally expect the level of energy applied to the risk process to fall to zero when the project completes. But a business does not usually perform just one project, and the same risk energetics cycle can be expected to occur on each project in the organisational portfolio. However, if the business is truly a learning organisation, one would expect to see a rising trend of energy and enthusiasm for risk management as one project gives way to the next, driven by the demonstrable success and value of managed risk on completed projects. Indeed, the presence of the factors already described should have a beneficial effect wider than just in each single project. If each project is exhibiting the internal factors of appropriate process, skilled facilitation and

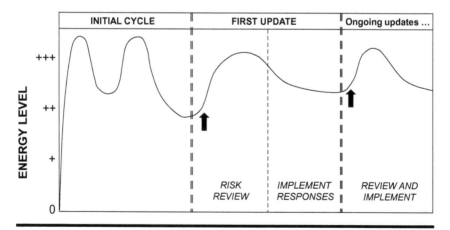

Figure 9.3   Risk Energetics – Updates and Reviews

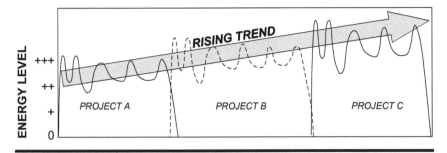

Figure 9.4   Risk Energetics – Rising Trend

adequate resourcing, and if the wider organisation is providing the right level of supporting infrastructure, and developing a risk-aware and risk-mature culture with visible senior management support, then the organisation should experience a growing maturity and effectiveness of risk management over time as it continues to learn. This will produce positive reinforcement and lead to increasing levels of attention, energy and enthusiasm for risk management. Figure 9.4 shows this trend, leading to a self-sustaining risk culture where the value of project risk management is recognised and expected.

Proving It Works

The concept of the CSF is that if it is present, it will promote success (in this case leading to enhanced risk management effectiveness), but if absent then success is hindered. However, when it comes to determining whether project risk management is in fact working, it is valid to ask how one would know. There is a philosophical problem with measuring risk management effectiveness: since risk is uncertain and may never happen, it is theoretically impossible to know the effect of any particular action on the outcome.

The expectation is that effective project risk management will lead to *fewer* threats turning into problems, and those that do will be *less severe.* Similarly, if risk management is working then *more* opportunities will be captured as benefits and savings within the project, at a rate that is *higher* than would be predicted by mere chance. A proper view of risk exposure will result in *appropriate* levels of contingency being set aside, *maximising* profit and margin for the project. And the original project plan will be *more* robust, and will be followed with *less* deviation and *reduced* volatility (assuming the plan takes proper account of risk).

There is however a problem with the preceding paragraph. Use of quantitative words such as 'fewer, more, less, maximised, reduced' implies two things:

1 An expectation of what would have happened without risk management (less or more than what?).
2 The ability to measure (or at least estimate) these variables.

How can we know how much risk is 'normal' on a given project, so that we can determine whether risk management is working? This is very difficult to quantify, and various metrics have been proposed, though none is perfect. Common risk process performance metrics include:

■ Active risks, closed threats, captured opportunities – both absolute numbers at a given point in time, and trend analysis;
■ Average risk-weighted score for threats and opportunities (using a Probability-Impact scoring system as described in Figure 3.3);
■ Quantitative Risk Analysis outputs, including percentage confidence levels for achieving project targets, trend graphs for risk reduction, and so on;
■ Planned risk responses and actions completed/outstanding/overdue;
■ Period between a risk being identified and resolved.

While these metrics can give some indication of whether risk management might be working on a particular project, it is better for an organisation that is serious about implementing risk management across all its projects to measure changes in overall project success rates with time. Simple measures such as those used by the Standish Group could be compiled and tracked within the organisation, determining the trend over time of how many projects succeed, fail or are challenged (as in Figure 2.2). Alternatively, more sophisticated and specific measures of project success can be constructed to assess key factors that are of importance to the particular business, such as the Three Ps (Predictability, Performance and Profit).

Why Bother?

Clearly, risk management is seen as a core part of the management of projects, which is why it has received increased and sustained attention. But having laid out the principles, process and psychology of project risk management in earlier chapters, we should conclude by challenging the real motives for doing it. Essentially, these fall under four headings:

1 *Mandated.* Many organisations and project teams include risk management as one of their project processes simply because they have been told to do so, either as a contractual or regulatory requirement, or in order to comply with internal company procedures. It is never a good idea just to do something merely to comply. This leads to lack of commitment to the risk process and a box-ticking mentality.

2 *Fear of failure.* Everyone working with projects knows that they are risky, and is looking for ways to minimise their risk exposure and maximise the chances of project success. But if the main motive for managing risk is to provide protection for when things go wrong, this will result in a very narrow focus for the risk process.

3 *Peer pressure.* Some organisations introduce project risk management because they see their competitors using it, or because it is viewed as a 'hot topic' or the latest management fad. Again, this motivation is flawed since the risk process requires sustained levels of commitment and energy, which cannot be motivated through comparison with others.

4 *Demonstrable benefits.* The only reason risk management should be used on projects (or in any other setting) is if it works, demonstrably and consistently. The risk process should deliver benefits to the project itself, and to project stakeholders including the project manager, team members, project sponsor, customers, suppliers and users. Since everyone working on projects is usually too busy, they will only do risk management if they see it working and helping them achieve their objectives. The ultimate measure of effective risk management is more successful projects with better project outcomes.

What benefits can be expected from implementing risk management on projects? Various studies have been published which list such benefits as those in Table 9.1, though these are mostly reported anecdotally from project risk practitioners (who might be thought to have a vested interest or at least a bias towards reporting benefits), and the data often lack quantitative credibility. Some hard empirical evidence for risk management effectiveness is beginning to emerge which is more credible and less anecdotal, linking good

Table 9.1 Benefits of Risk Management (From Hillson & Simon, 2020)

Generic benefits of risk management	
'Hard' benefits	*'Soft' benefits*
Enables better informed and more believable plans, schedules and budgets	Improves corporate experience and general communication
Increases the likelihood of a project adhering to its schedules and budgets	Leads to a common understanding and improved team spirit
Leads to the use of the most suitable type of contract	Helps distinguish between good luck/good management and bad luck/bad management
Allows a more meaningful assessment of contingencies	Helps develop the ability of staff to assess risks
Discourages the acceptance of financially unsound projects	Focuses project management attention on the real and most important issues
Contributes to the build-up of statistical information to assist in better management of future projects	Facilitates greater risk-taking, this increasing the benefits gained
Enables a more objective comparison of alternatives	Demonstrates a responsible approach to customers
Identifies, and allocates responsibility to, the best risk owner	Provides a fresh view of the personnel issues in a project
Organisational benefits of risk management	
Compliance with corporate governance requirements	Better reputation as a result of fewer headline project failures
A greater potential for future business with existing customers	Better customer relations due to improved performance on current projects
Reduced cost base	A less stressful working environment

project risk management practice to better project outcomes, but this is currently not widely available in published literature.

Rather than simply listing potential benefits which have been derived anecdotally, it would be preferable if it were possible to perform some kind

Table 9.2   Cost-Benefit Analysis for Project Risk Management

Short-term cost-benefit analysis for project risk management	
Costs	*Benefits*
▪ Provide risk infrastructure (training, tools and so on) ▪ Provide resources for the risk process ▪ Implement agreed risk responses	▪ Improved project predictability ▪ Successful project delivery ▪ Enhanced customer satisfaction
Long-term cost-benefit analysis for project risk management	
Costs	*Benefits*
▪ Commitment ▪ Consistency ▪ Continuation ▪ Culture	▪ Business growth ▪ Team motivation ▪ Fewer surprises ▪ Enhanced reputation

of cost-benefit analysis for project risk management. This can be done at two levels: short-term and long-term. The items listed in Table 9.2 are wider than the elements of a traditional cost-benefit analysis. They do however indicate the types of investment which an organisation must make if it is serious about implementing risk management on projects, as well as the broader benefits that can be obtained.

Obviously, the expectation is that the degree of benefits will exceed the costs deployed, thus justifying the use of risk management in projects as an effective approach to dealing with their intrinsic uncertainty. If these costs and benefits are measured and reported consistently and openly, they will also enable project managers within a business to support use of risk management on their projects, selling it to senior management, and encouraging the organisation to invest in the CSFs described above.

And Finally...

We set out in this book to discover why risk management is important in the context of projects, how it should be implemented, how risk outputs should be used both within and outside the project, and what is necessary to maximise risk management effectiveness. Although we have described a generic project risk process, risk management is so much more than the Three Ts of

Tools, Techniques and Training. It includes the softer elements of human behaviour and psychology, which must be understood if the results of the risk process are to be used properly to support good decision-making within projects. It is also much wider than just projects, informing the way the broader organisation operates and is managed.

Risk management is an essential contributor to project and business success, because of its relentless focus on finding and managing those factors that affect achievement of objectives. Done properly, it is one of the most powerful weapons in the project manager's armoury, defending against the worst effects of inevitable uncertainty while allowing the project and the business to create advantage and innovation. Risk management is truly one of the fundamentals of project management. Projects that fail or are challenged reach that position as a direct result of the consequences of unmanaged risk. Successful projects are the ones which understand the risks they face and deal with them effectively.

References and Further Reading

Association for Project Management. (2008). *Prioritising Project Risks*. High Wycombe, Bucks, UK: Association for Project Management.

Association for Project Management. (2019). *APM Body of Knowledge, 7th edition*. Princes Risborough, Bucks, UK: APM Publishing.

Association for Project Management. (2024). *Project Risk Analysis & Management (PRAM) Guide, third edition*. Princes Risborough, Bucks, UK: APM Publishing.

Axelos. (2022). *M_o_R 4: Management of Risk: Creating and protecting value, 4th edition*. London, UK: PeopleCert International.

Barber, R. B. (2003). *A Systems Toolbox for Risk Management*. Proceedings of ANZSYS Conference: Monash, Australia, November 2003. Available online: https://www.anzsys.org/conferences/anzsys03/der3000059_2.pdf

Barber, R. B. (2023). *Risk Leadership in Complexity*, in *The Risk Management Handbook, second edition*, edited by D. A. Hillson. London, UK: Kogan Page Publishers.

Belack, C., Di Filippo, D. & Di Filippo, I. (2019). *Cognitive Readiness in Project Teams: Reducing Project Complexity and Increasing Success in Project Management*. Abingdon, Oxon, UK: Routledge.

Cooper, D. F., Bosnich, P., Grey, S., Purdy, G., Raymond, G., Walker, P. & Wood, M. (2014). *Project Risk Management Guidelines: Managing Risk with ISO 31000 and IEC 62198, second Edition*. Chichester, UK: J Wiley.

Coyle, R.G. (1997). *The nature and value of futures studies, or do futures have a future?* Futures, Vol. 29(1), pp. 77–93.

European Commission Centre of Excellence in Project Management (CoEPM²). (2018). *PM² Project Management Methodology – Guide 3.0*. Brussels, Belgium: European Commission.

Government Office for Science (2021). *A brief guide to futures thinking and foresight*. London, UK: Crown Copyright. Available online: https://www.gov.uk/government/publications/futures-thinking-and-foresight-a-brief-guide

Hancock, D. (2010). *Tame, Messy & Wicked Risk Leadership*. Farnham, UK: Gower.

Hancock, D. (2012). The case for risk leadership. *Strategic Risk Europe*, 1 November 2012. Available online: https://www.strategic-risk-europe.com/the-case-for-risk-leadership/1399431.article

HarperCollins (n.d.). London, UK: HarperCollins. Available online: www.collinsdictionary.com/dictionary/english

Hass, K.B. (2009). *Managing Project Complexity: A new model*. Vienna, VA: Management Concepts.

Hillson, D. A. (2003). Using a Risk Breakdown Structure in Project Management. *Journal of Facilities Management, 2*(1), 85–97.

Hillson, D. A. (2011). *Enterprise Risk Management: Managing uncertainty and minimising surprise*, in *Advising Upwards: A framework for understanding and engaging senior management stakeholders*, ed. L. Bourne. Farnham, UK: Gower.

Hillson, D. A. (2013). *The A-B-C of Risk Culture: How to be Risk-Mature*. Proceedings of PMI Global Congress North America 2013, New Orleans, LA, USA, October 2013

Hillson, D. A. (2019). *Capturing Upside Risk: Finding and managing opportunities in projects*. Boca Raton, FL, USA: Taylor & Francis.

Hillson, D. A. (2022). *Taming the Risk Hurricane: Preparing for major business disruption*. Oakland, CA, USA: Berrett-Koehler Publishers.

Hillson, D. A. (2023a). *Portfolio Risk Management*, in *Strategic Portfolio Management in* the Multi-Project and Program Organisation, eds. K. Angliss & P. Harpum. Abingdon, UK: Routledge.

Hillson, D. A. (ed.) (2023b). *The Risk Management Handbook: A practical guide to managing the multiple dimensions of risk, second edition*. London, UK: Kogan Page Publishing.

Hillson, D. A. & Murray-Webster, R. (2007). *Understanding and Managing Risk Attitude, second edition*. Aldershot, UK: Gower.

Hillson, D. A. & Simon, P. W. (2020). *Practical project risk management: The ATOM Methodology, third edition*. Oakland, CA, USA: Berrett-Koehler Publishers.

Hulett, D. T. (2009). *Practical Schedule Risk Analysis*. Farnham, UK: Gower.

Hulett, D. T. (2011). *Integrated Cost–Schedule Risk Analysis*. Farnham, UK: Gower.

Institute of Risk Management (2018). *Horizon Scanning: A practitioner's guide*. London, UK: Institute of Risk Management

Institution of Civil Engineers, Institute and Faculty of Actuaries. (2014). *Risk Analysis & Management for Projects (RAMP), third edition*. London UK: ICE Publishing.

International Centre for Complex Project Management (2021). *Harnessing Emergence in Complex Projects: Rethinking risk, opportunity and resilience*. Deakin West, ACT, Australia: International Centre for Complex Project Management (ICCPM).

International Electrotechnical Commission (2024). *IEC 62198:2024: Managing risk in projects – Application guidelines*. Geneva, Switzerland: International Electrotechnical Commission.

International Organization for Standardization. (2018). *ISO 31000:2018: Risk Management Guidelines*. Geneva, Switzerland: International Organization for Standardization.

International Organization for Standardization. (2021). *ISO 21500:2021: Project, programme and portfolio management – Context and concepts.* Geneva, Switzerland: International Organization for Standardization.

International Project Management Association. (2015). *Individual Competence Baseline for Project Management*, version 4.0. Nijkerk, The Netherlands: International Project Management Association (IPMA).

Knight, F. H. (1921). *Risk, Uncertainty and Profit.* New York, NY, USA: Houghton Mifflin.

Markowitz, H. (1959). *Portfolio Selection: Efficient Diversification of Investments.* New York, NY, USA: J Wiley.

Murray-Webster, R. & Hillson, D. A. (2008). *Managing Group Risk Attitude.* Aldershot, UK: Gower.

Murray-Webster, R. & Hillson, D. A. (2021). *Making Risky and Important Decisions: A leader's guide.* Boca Raton, FL, USA: Taylor & Francis.

Murray-Webster, R. & Pullan, P. (2023). *Making Risk Management Work: Engaging people to identify, own and manage risk, second edition.* Abingdon, Oxon, UK: Routledge.

Obeng, E. (1997). *New Rules for the New World: Cautionary Tales for the New World Manager.* Oxford, UK: Capstone Publishing.

Project Management Institute. (2017a). *PMI Lexicon of Project Management Terms (version 3.2).* Newtown Square, PA, USA: Project Management Institute.

Project Management Institute. (2017b). *A Guide to the Project Management Body of Knowledge PMBOK® Guide – Sixth Edition.* Newtown Square, PA, USA: Project Management Institute

Project Management Institute. (2019). *The Standard for Risk Management in Portfolios, Programs, and Projects.* Newtown Square, PA, USA: Project Management Institute.

Rhyne, R. (1974). Technological forecasting within alternative whole futures projections. *Technological Forecasting and Social Change*, Volume 6, pp. 133–162.

Seville, E. (2016). *Resilient organizations: How to survive, thrive and create opportunities through crisis and change.* London, UK: Kogan Page Publishers.

Stauffer, M., Seifert, K., Aristizábal, A., Chaudhry, H. T., Kohler, K., Hussein, S. N., Leyva, C. S., Gebert, A., Arbeid, J., Estier, M., Matinyi, S., Hausenloy, J., Kaur, J., Rath, S., & Wu, Y-H. (2023) *Existential Risk and Rapid Technological Change: Advancing risk-informed development.* Geneva, Switzerland: United Nations Office for Disaster Risk Reduction (UNDRR). Available online: https://www.undrr.org/publication/thematic-study-existential-risk-and-rapid-technological-change-advancing-risk-informed

Strella, K., Riedel, J. & Gerhardt, T. (2015) *Potential: The Raw Material of the Future.* Talent Quarterly #6, May 2015, pp. 37–42. Available online at https://www.egonzehnder.com/what-we-do/executive-search/insights/potential-the-raw-material-of-the-future

The Standish Group. (2015). CHAOS Report 2015. The Standish Group International. Available online: https://www.standishgroup.com/sample_research_files/CHAOSReport2015-Final.pdf (accessed 1 May 2023)

Williams, T. M. (2002). *Modelling Complex Projects.* Chichester, UK: J Wiley.

Index

Note: Page numbers in *italics* indicate figures, and page numbers in **bold** indicate tables in the text.

A-B-C Model (attitude-behaviour-culture) 118, *118*
action owner 51, **57**
adaptive resilience 68-70
Association for Project Management (APM) 21–2, 32, 64
ATOM Risk Methodology 22

bubble diagram 43, *43*
business continuity management (BCM) 69, 70

cognitive readiness 71–2
complex project: characteristics 64; definition 62; risk management 63, 68–70
complexity, definition, 61–2
complicated, definition, 61–2
corporate risk champion 135
corporate risk sponsor 135
critical success factor (CSF): characteristics 128–9, 137; external factors 132–5; internal factors 130–2; risk management as CSF 23–4
Cynefin framework 60–3, *60*

decision trees 45
decision-making 83–6, *84*

enterprise risk management (ERM) 103–4, 107–8
existential risk 3

facilitation 131–2
futures thinking 63, 65–8, *67*; field anomaly relaxation (FAR) 66; horizon scanning 66; scenario analysis 65; trend analysis 65

IEC62198:2024 29, 32
individual risk 20–2, **22**
Institute of Risk Management (IRM) 74
International Centre for Complex Project Management (ICCPM) 59–60, 73–4
ISO31000:2018 8, 9, 32

leadership and management 115, **115**

objectives 11, 26, 100; hierarchy 101, *101*

old world/new world 4–5, *4*
overall risk 20–2, **22**, 45; sources 39

permacrisis 3
polycrisis 3
probability and impacts: defining terms
    34–5, **35**
probability-impact matrix 41–2, *42*
probability-impact scores 41–2, *42*
project: characteristics 16–17; definition
    13–14, **14**; design 17–18;
    external environment 18–19
project continuity management
    69, 70–3
project lifecycle 90–6, *91*, 135–7;
    baseline estimating 97; change
    control 98; conception 92;
    contingency management
    97; contract negotiation and
    procurement 96; initiation
    93; iterative 96; resource
    allocation 97; selection of
    development options 97;
    viability 92; waterfall 95
project management 21–2, 32; response
    to uncertainty 22–3;
Project Management Institute (PMI)
    21, 22, 32
project mindset 113
prompt list 38

risk: and projects 16–17; and reward
    18–19, *19*; definition 7–8;
    hierarchy 102; threat and
    opportunity 8–11; uncertainty
    that matters 7–8, **9**
risk appetite 84, 86–8, 116, 121–2
risk attitude: and risk process 81–3, **82**;
    definition 76, **79**; risk-averse
    77–9, **79**; risk-neutral 78–9,
    79; risk-seeking 78–9, **79**;
    risk-tolerant 78–9, **79**

risk attitude spectrum 77–8, *78*
Risk Breakdown Structure (RBS) 34,
    36, 38, 43
risk capacity 85
risk champion 87; Corporate Risk
    Champion 135; as facilitator
    131–2; risk roles and
    responsibilities **57**
risk culture 116, 117–18; definition 117;
    organisational 134
risk dashboard 52
risk efficiency 109–111, 112; risk-
    efficient frontier 110–111
risk efficiency graph 110–11, *111*
risk energetics 128–37, *128, 130,
    136, 137*; external factors
    132–5; internal
    factors 130–2
risk leadership 114–26; behaviours
    123–5; influencing others
    122–3; responsibilities
    115–17, *116*; Risk Leadership
    Network 117
risk management: across boundaries
    104–6; benefits 139–141, **140**;
    and business performance
    100; cost-benefit analysis
    140–1, **141**; as CSF 23–4,
    24; infrastructure 133;
    management support 134–5;
    motivation 139; programmes
    and portfolios 109–112;
    resources 132
Risk Management Plan 35, **37**, 38, 39,
    41, 58
risk management standards 32
risk metalanguage 38–9
risk mindset 118–19
risk owner 39, 48, 51, 57, 106
risk policy 116, 120–1
risk prioritisation chart 43–4, *44*
risk prioritisation factors 42, 50

risk process 116, 122; design, 131; generic 29–31, *31*, **32**; informal 26–8, 30, **30**; international standards **32**; intuitive 26; metrics 138; Monte Carlo simulation 45–7; Post–Project Risk Review 53–4; Qualitative Risk Assessment 40–4; Quantitative Risk Analysis 45–8; Risk Communication 52; Risk Identification 35, 37–40; Risk Management Plan 35, 37, 41; Risk Process Initiation 33–5; risk reports 52; Risk Response Implementation 51–2; Risk Response Planning 48–51; risk response strategy 49–50; Risk Review 52–3; scalable 54–5; S-curve 47–8, *47*, *48*

risk register 39–40, **40**, 43, 51, 53

risk roles and responsibilities **56–7**; action owner **57**; project manager **56**; project sponsor **56**; risk champion **57**; risk owner **57**; stakeholders **57**; team members **57**

risk threshold 33–4, 84–5, 110, 112, 116, 121–2

risk vision 116, 120–1

secondary risk 50, 52

Seven A's model 86–8, *86*

stakeholders 17; and resilience 73; communication to 52, 108; risk tolerance/threshold/appetite 26, 34, 92, 112

Standish Group CHAOS Report 14–15, *15*

system dynamics 64

Three Ts (tools, techniques, training) 75

triple strand 80–1, *80*

uncertainty 1; and risk 5–6, **6**

United Nations Office for Disaster Risk Reduction (UNDRR) 3

VUCA 59

WEIRD nations 1

Milton Keynes UK
Ingram Content Group UK Ltd.
UKHW030903141024
449569UK00026B/1313